18 VIIIA
		13 IIIA	14 IVA	15 VA	16 VIA	17 VIIA	2 **He** Helium 4.0026
		5 **B** Boron 10.81	6 **C** Carbon 12.011	7 **N** Nitrogen 14.007	8 **O** Oxygen 15.999	9 **F** Fluorine 18.998	10 **Ne** Neon 20.180
11 IB	12 IIB	13 **Al** Aluminium 26.982	14 **Si** Silicon 28.085	15 **P** Phosphorus 30.974	16 **S** Sulfur 32.06	17 **Cl** Chlorine 35.45	18 **Ar** Argon 39.948
29 **Cu** Copper 63.546	30 **Zn** Zinc 65.38	31 **Ga** Gallium 69.723	32 **Ge** Germanium 72.630	33 **As** Arsenic 74.922	34 **Se** Selenium 78.971	35 **Br** Bromine 79.904	36 **Kr** Krypton 83.798
47 **Ag** Silver 107.87	48 **Cd** Cadmium 112.41	49 **In** Indium 114.82	50 **Sn** Tin 118.71	51 **Sb** Antimony 121.76	52 **Te** Tellurium 127.60	53 **I** Iodine 126.90	54 **Xe** Xenon 131.29
79 **Au** Gold 196.97	80 **Hg** Mercury 200.59	81 **Tl** Thallium 204.38	82 **Pb** Lead 207.2	83 **Bi** Bismuth 208.98	84 **Po** Polonium (209)	85 **At** Astatine (210)	86 **Rn** Radon (222)
111 **Rg** Roentgenium (282)	112 **Cn** Copernicium (285)	113 **Nh** Nihonium (286)	114 **Fl** Flerovium (289)	115 **Mc** Moscovium (290)	116 **Lv** Livermorium (293)	117 **Ts** Tennessine (294)	118 **Og** Oganesson (294)

65 **Tb** Terbium 158.93	66 **Dy** Dysprosium 162.50	67 **Ho** Holmium 164.93	68 **Er** Erbium 167.26	69 **Tm** Thulium 168.93	70 **Yb** Ytterbium 173.05	71 **Lu** Lutetium 174.97
97 **Bk** Berkelium (247)	98 **Cf** Californium (251)	99 **Es** Einsteinium (252)	100 **Fm** Fermium (257)	101 **Md** Mendelevium (258)	102 **No** Nobelium (259)	103 **Lr** Lawrencium (266)

ELEMENTARY

ELEMENTARY
The Periodic Table Explained

JAMES M. RUSSELL

Michael O'Mara Books Limited

First published in Great Britain in 2019 by
Michael O'Mara Books Limited
9 Lion Yard
Tremadoc Road
London SW4 7NQ

A CIP catalogue record for this book is available from the British Library.

Papers used by Michael O'Mara Books Limited are natural, recyclable
products made from wood grown in sustainable forests.
The manufacturing processes conform to the environmental
regulations of the country of origin.

ISBN: 978-1-78929-102-5 in hardback print format
ISBN: 978-1-78929-103-2 in ebook format

1 2 3 4 5 6 7 8 9 10

Cover design: Ana Bjezancevic
Designed and typeset by E-type

Printed and bound by CPI Group (UK) Ltd, Croydon, CR0 4YY

www.mombooks.com

Contents

Introduction:
Mendeleev's Brilliant Idea

The periodic table is one of the most transformative scientific discoveries of the last two centuries, yet its inception required no scientific instruments or experiments – just a pen, a piece of paper and a talented Russian chemist, Dmitri Mendeleev (1834–1907). In the early 1860s, fascinated by atomic theory – the idea that elements are uniquely defined by their atomic make-up – Mendeleev wanted to explore the idea of organizing all of the known elements in a simple diagram.

At the time, it was known that matter was made up of 'elements', sixty-two of which had been identified. Mendeleev started by arranging them in order of their atomic mass number, which is the total number of neutrons and protons in an atom of that element. (The nucleus of an atom is made up of protons and neutrons, around which a cloud of electrons orbits: the electrons are so light that their mass is ignored in calculating atomic mass.)

At first, he simply laid out the elements in a long row. However, the crucial insight came when he realized that within this row there were patterns: elements with similar properties were appearing at specific 'periods' within it.

By cutting up the row and rearranging it in several shorter rows so that similar elements were above each other in columns, he came up with the first version of the periodic table. His left-hand column included sodium, lithium and potassium – these are all solids at room temperature (which is usually taken to mean about 20°C), they all tarnish easily and all react vigorously when mixed with water.

Mendeleev also came up with the 'periodic law', a summary of his insight that the elements fall into recurring groups, meaning that elements with similar properties occur at regular intervals. The qualities that constitute 'similar properties' include their electronegativity, ionization energy, metallic character and reactivity of the elements.

As he continued to work on the table, which he first published in 1869, he occasionally tweaked the array, and found that the patterns were reinforced if he occasionally broke his own rules, by placing some elements out of order and leaving gaps. For instance, arsenic in the original table was in period 4 group 13, but Mendeleev believed it fitted more closely with the elements in group 15, so he moved it to that position, leaving groups 13 and 14 on that row empty.

The brilliance of this decision would later be vindicated when gallium and germanium were discovered; elements that fit perfectly into the empty spaces before arsenic. Over the following 150 years, more elements have been found or synthesized: argon, boron, neon, polonium, radon and many more besides. And each of these have been slotted into a spot on the periodic table, which currently contains 118 elements.

Mendeleev's rearrangement of the table was intuitive, based on the properties of the elements, but in his lifetime the table continued to be ordered by atomic mass. It was only in 1913 that Henry Moseley proved that the underlying principle of the elements' order was not atomic mass after all, but the slightly different quality of 'atomic number'. This is determined solely by the number of protons in an atom. Protons carry a positive electric charge, so the atomic number is a pure measure of the positive charge on the nucleus: it has since been discovered that the number of negatively charged electrons around the nucleus is equal to the number of protons, making the net charge of a normal atom equal to zero. And Moseley's findings led to the discovery of further elements, as the newly reorganized table had further gaps in it (see page 134 for more detail).

It is now well established that any element can be uniquely identified by its atomic number. But the number of neutrons is still significant, as it defines different 'isotopes'. For instance, any atom with a single proton is a hydrogen, but while an ordinary atom of hydrogen has no neutrons (and can also be referred to as protium or ^1H), there are two further naturally occurring isotopes: deuterium (^2H), which has one proton and one neutron; and tritium (^3H), which has one proton and two neutrons. And it is possible to synthesize further isotopes; if you bombard tritium with deuterium nuclei, you can make hydrogen-4 (^4H), which has one proton and three neutrons. However, this is a highly unstable isotope, which will rapidly decay back into one of the naturally occurring isotopes.

Mendeleev's humble table not only predicted undiscovered substances: it also led chemists to a deeper understanding of atoms themselves. Chemists would eventually come to understand that the similarity in groups or columns of the periodic table was defined by the subatomic structure of the element. Electrons in an atom are arrayed in a number of levels, which are known as shells. Each of these has a limited number of spaces: the first level has just two spaces, then each of the next two levels has eight spaces.

As the atomic number increases, these spaces are gradually 'filled up'. Elements in a particular group of the table have the same number of shells in their outer (or 'valence') shell – and it is the number and arrangement of electrons in this shell that dictates how that atom will behave in a chemical reaction, in which different atoms exchange electrons and the molecules consisting of those atoms are transformed in the process. Elements that have a full outer shell (including the noble gases, which include helium, neon and argon) are more stable and less reactive, while elements with spaces on the outer shell are more reactive.

It is also important to know that the specific arrangement of the same number of electrons can lead to differences in the way that atoms of an element bond to each other; we'll see how different bonding structures of carbon lead to the quite different substances diamond, graphite and soot, which are known as 'allotropes' of carbon.

So, our current understanding of the chemical structure of the universe is based firmly on Mendeleev's periodic table. Its use as a theoretical tool was one of the keys that unlocked the astonishing micro-world of subatomic particles. But this breakthrough was only made possible by the development of atomic theory, which had become widely accepted through the nineteenth century.

John Dalton was a gifted early nineteenth-century amateur scientist. He was a dissenter, so was barred from most British universities but was educated by the blind philosopher John Gough. After having to leave the radical 'New College' in Manchester for financial reasons, he carried on with his own experiments and contributed significantly to our knowledge of weather prediction, how gases behave and colour blindness.

However, his most significant legacy was his statement of what came to be known as 'atomic theory'. While pondering the fact that elements combine in predictable and regular ways with each other (for instance, compounds separate into definite proportions of their constituent elements), he came up with the first set of 'atomic weights'. In 1810, he published a list of the atomic weights of hydrogen, oxygen, nitrogen, carbon, sulphur and phosphorus.

It was this insight – that individual atoms of a given element are identical and have a definable mass – that underpinned progress in chemistry in subsequent decades and led on to Mendeleev's periodic table.

So, now that we have taken an overview of how atomic theory and the periodic table were developed and their significance, let's take a whistle-stop tour of the 118 known elements, in order of their atomic number.

Elements 1-56

Hydrogen

Category: non-metal
Atomic number: 1
Colour: colourless
Melting point: −259°C (−434°F)

Boiling point: −253°C (−423°F)
First identified: 1766

Hydrogen is the simplest possible atom, with a nucleus of only one proton and one electron. It was one of the first elements to be formed after the Big Bang, and remains the most abundant in the universe – even though it has been burning in countless stars, where it is fused into helium, it still makes up more than 75 per cent of the detectable universe and appears in more compounds than any other element.

A light, colourless, highly flammable gas, it is rich on our planet in the form of water (two hydrogen atoms bonded to one oxygen atom). The weak bonds that hydrogen forms in molecules give water its relatively high boiling point, allowing it to exist in liquid form in the Earth's atmosphere, while at low temperatures, the hydrogen bonds adjust and hold the oxygen atoms apart in a kind of crystal lattice: most substances are denser in their solid state than in their liquid state, but this lattice makes ice lighter than water, which is why icebergs float.

Hydrogen also bonds with carbon to form hydrocarbons, including fossil fuels such as coal, crude oil and natural gas (it is a highly combustible element – when you see a candle burning, this is mostly because hydrogen is released from the oil or tallow and burns when it comes into contact with oxygen). Without hydrogen, we wouldn't have the heat and light from the constant nuclear fusion of the sun.

The sixteenth-century alchemist Paracelsus was the first to observe the phenomenon that bubbles of a flammable gas are produced when metal is mixed with strong acids. (Chemistry teachers use the mnemonic MASH, to remind students that metals + acids produce salts and hydrogen.) In 1671, Robert Boyle observed the same thing when iron filings were mixed with hydrochloric acid (a compound of hydrogen and chlorine). It was nearly a century later, in 1766, that Henry Cavendish realized this gas was a separate element, though he called it inflammable air, which he wrongly identified as phlogiston. In 1781, when he found that this gas produced water when it was burned, Cavendish suggested that the oxygen it was combining with was 'dephlogisticated air'. It took the brilliant French chemist Antoine Lavoisier, in 1783, to give hydrogen its current name, which is derived from the Greek for 'water producer'.

Phlogiston, a Dead Horse

The phlogiston theory, which misled Cavendish, was the now-deceased idea that all combustible bodies contained a fire-like element (named from the Ancient Greek word for 'flame'). The theory was that substances containing phlogiston became dephlogisticated when they burned. The first cracks in this theory came when it was shown that some metals gained weight rather than losing it when they burned, and Lavoisier more or less disproved it when he used closed vessel experiments to show that combustion requires a gas (oxygen) that has a measurable mass.

Hydrogen is extremely light, one reason why it isn't commonly found in pure form in the air (it basically just floats away and can escape the atmosphere). It is much lighter than oxygen or nitrogen, which is why it was the first gas used to fill a hot-air balloon. It would also be used in airships (hot-air balloons with a rigid structure) – but the boom in airship (or zeppelin) travel in the early twentieth century came to an abrupt end after the spectacular crash of the passenger airship *LZ 129 Hindenburg* in 1937.

Hydrogen is used, however, in some NASA rockets, including the main Space Shuttle engines, which are powered by burning liquid hydrogen and pure oxygen. And it could be the clean fuel of the future, replacing fossil fuels in cars, either directly or, more likely, in the form of fuel cells, where it would produce only water vapour as a waste product. There are problems to overcome, though: mass storage of such a highly flammable substance would be risky, and hydrogen is either refined from hydrocarbons, which produces more greenhouse gases, or through electrolysis of water, powered by electricity, which will most likely have been produced using fossil fuels in the first place.

There are many other uses for hydrogen: to produce ammonia for fertilizers, to create compounds such as cyclohexane and methanol (which are used in the production of plastics and medicines), and in the manufacture of margarine, glass and silicon chips, among other important products.

Helium

Category: noble gas

Atomic number 2

Colour: colourless

Melting point: −272°C (−458°F)

Boiling point: −269°C (−452°F)

First identified: 1895

Nearly everything in the universe that isn't hydrogen is helium – all the other elements make up only about 2 per cent of the mass of the universe, in spite of being heavier than the two lightest, simplest elements.

However, helium is not that common on Earth – indeed, it wasn't until 1895 that we were sure it existed here at all.

As one of the noble gases, helium is the second least reactive element, so, unlike hydrogen, it isn't extensively captured in compounds. However, like hydrogen, its pure form is lighter than air and prone to escaping the Earth's atmosphere. We find helium as part of natural gas beneath the ground, where it has been formed during the decay of radioactive elements like thorium and uranium.

Helium does make up about 24 per cent of the mass of the sun: in the extreme temperatures of the star, hydrogen nuclei undergo a process of fusion and helium is formed. This creates huge amounts of energy, and possibly an inexhaustible and environmentally friendly solution to our future energy requirements, although we are probably decades away from recreating the nuclear fusion process on the Earth.

One way to identify elements is by using a spectroscope, an instrument that analyses the differently coloured flames created by different elements to generate a kind of 'elemental fingerprint' in which the light is split into a series of coloured lines, rather than a continuous spectrum. During a solar eclipse in 1868, two separate astronomers (Jules Janssen from France and the English Norman Lockyer) noticed that there were some clear lines in the spectrum of the sun that didn't match any known element. Lockyer suggested this was an undiscovered element and named it helium after the Greek sun god Helios. The '-ium' ending shows that he assumed it was a kind of metal, as this is the only non-metal element which takes that ending. Over subsequent decades there was no further evidence of the existence of helium, though Lockyer was vindicated in 1895 when chemist William Ramsay found traces of helium in the gas given off by a chunk of uranium that had been treated in acid; the helium had already formed within the rock but was released as the acid dissolved part of the surface.

The Mickey Mouse Effect

After helium was found in American natural gas deposits, the first helium production plant (which supplied the army with gas for its barrage balloons) opened in 1915 in Texas. From 1919, the US navy experimented with gas mixes to combat the problem of nitrogen narcosis in deep-sea divers. In the notes from a 1925 experiment, divers breathing a helium and oxygen mixture complained that the change in vocal characteristics had made communication difficult. (The comical squeakiness is caused by soundwaves, which travel faster in any gas that is lighter than air.) Over time, helium was produced more extensively and used to fill party balloons, at which point a new generation of children learned the same trick for themselves.

Helium has the lowest boiling point of any substance and can be used in supercooling other substances; it's used, for instance, in the Large Hadron Collider, in superconducting magnets such as those in MRI scanners, and to cool the liquid hydrogen used in some NASA rockets. (Supercooling is the process of reducing the temperature of a substance to below its freezing point without it turning into a solid.) Some car airbags contain helium, because it diffuses so rapidly when decompressed (although nitrogen and argon are also used for this purpose).

There is reason to worry about our supply of helium. The market price has crashed since the 1990s' privatization of US stocks, but it is a limited resource, which is only replenished very slowly within the planet. So, while helium party balloons are fun, bear in mind that the helium from them can escape and leave the atmosphere, so they may not be the smartest way to use this particular noble gas.

Lithium

Category: alkali metal
Atomic number: 3
Colour: silvery white
Melting point: 181°C (358°F)

Boiling point: 1,342°C (2,448°F)
First identified: 1817

Scientists believe that the only element initially created in the Big Bang other than hydrogen and helium was the metal lithium, albeit in much smaller quantities. Lithium was first found in 1800 in petalite, a pale or transparent ore that can be formed into gemstones, but it was only in 1817 that the chemist Johan August Arfwedson realized that this ore contained a previously unknown element. He named it after the Greek word *lithos*, which means stone, because it had been found in an ore, whereas other alkali metals such as potassium and sodium were first clearly identified in organic matter, such as plant ashes and animal blood. In 1821, William Thomas Brande used electrolysis of lithium oxide to isolate pure lithium. It is soft and silvery, with the lowest density of any metal, and has a violent reaction with water.

Lithium doesn't naturally occur as a metal on Earth, as it is so reactive that the pure form has to be kept under a protective oil to protect it from corrosion. Instead, it is found in trace amounts in various igneous rocks and dissolved in the water of mineral springs.

The Strange Case of 7 Up

Most people know that Coca-Cola originally contained cocaine, but did you know that the original version of 7 Up included lithium citrate, which can be used to medicate mood swings? The Howdy Corporation, founded by Charles Leiper Grigg, launched a new soft drink in 1920, which was called 'Bib-Label Lithiated Lemon-Lime Soda'. The name was eventually changed to 7 Up but it became illegal for beverage manufacturers to include lithium in their products in 1948.

The physician, Soranus of Ephesus, may have been unknowingly using lithium as a medicine in the second century when he prescribed the alkaline water from a local spring (which is now known to contain lithium) for mania and melancholia. Lithium carbonate has been used for medicinal purposes since the nineteenth century, with varying success; in particular, since the 1940s, it has been used as a treatment for bipolar disorder, in spite of some controversy over its side effects and potential toxicity.

Lithium can also be used in alloys with aluminium and magnesium, making them stronger and lighter – these are used in aircraft, bicycles and trains, where a lighter metal allows for greater speed. And a lithium compound is used for the cathodes in lithium batteries, which are longer lasting than most standard battery types.

Beryllium

Category: alkaline earth metal

Atomic number: 4

Colour: silvery white

Melting point: 1,287°C (2,349°F)

Boiling point: 2,469°C (4,476°F)

First identified: 1798

Before humanity knew about beryllium, we were fascinated by a mineral that contains this rare metal: beryl (whose scientific name is beryllium aluminium silicate) forms into a variety of beautiful gemstones, including aquamarine, heliodor and emerald (the attractive green colour of which is caused by small quantities of chromium or vanadium). The ancient Egyptians, Celts and Romans all placed a high value on emeralds, which at that time came from deposits in central Europe or the Indian subcontinent, although there were later discoveries in South America and Africa.

A Dangerous Metal

Early fluorescent lights had a coating of chemicals, including beryllium oxide; unfortunately, beryllium fumes are toxic and cause an inflammatory lung condition called berylliosis. This manufacturing method was discontinued in the late 1940s after an outbreak of the disease at an American plant that made the lamps. The nuclear physicist Herbert L. Anderson, who contributed significantly to the Manhattan Project, died in 1988 after a four-decade struggle with the disease, which he contracted while working with uranium in the nuclear programme.

In the eighteenth century, the French priest and mineralogist René-Just Haüy asked the chemist Louis Nicolas Vauquelin whether beryl might have an unknown element in its chemical make-up. In 1798, Vauquelin identified the presence of a new metal, which he announced, calling it glaucinium from the Greek word for sweet (*glykys*) due to the sweetish taste of some of its compounds. However, the name beryllium, derived from beryl, became more widespread and stuck.

It took thirty years of experiments before French and German chemists independently managed to isolate beryllium, by extracting it from beryllium chloride through a reaction with potassium. It is a soft, silvery-white metal with a low density. Not at all common in the universe, it was formed after the Big Bang, and can't be formed in stars, only in supernova explosions.

Beryllium has some peculiar qualities – James Chadwick won the Nobel Prize in 1935 for his discovery that it reflects neutrons while it is actually transparent to X-rays. These qualities explain some of its uses today – beryllium foil is used in X-ray lithography and the metal is employed in X-ray tube windows, within space telescopes and inside nuclear warheads, where it is used to reflect the neutrons that are bombarding uranium. It is also used in alloys with copper or nickel (whose electrical and thermal conductivity are thereby increased) for items such as gyroscopes, electrodes and springs, and in other alloys that are used in aircraft and satellites.

Boron

Category: metalloid
Atomic number: 5
Colour: variable
Melting point: 2,076°C (3,769°F)

Boiling point: 3,927°C (7,101°F)
First identified: 1732

As with beryllium, boron was known about for centuries only through one of its compounds, borax (which also answers to the names of sodium borate, sodium tetraborate and disodium tetraborate). Borax is a salt of boric acid, and comes in the form of white, soft crystals that dissolve when you put them in water. Historically, it has been used as a detergent, cosmetic, fire retardant, insect repellent, and by ancient goldsmiths as a flux: a substance added to metal to make it easier to work with. It was mentioned in Chaucer's *Canterbury Tales* and was used in Elizabethan England as a cosmetic, mixed with egg shells and oil and smeared on the face to create the white foundation that was popular at the time.

In the medieval period, the only source of borax was crystallized deposits from a lake in Tibet – it was traded along the Silk Road to the Arabian Peninsula and eventually Europe. It became more widely used in the nineteenth century after further deposits were discovered, notably in the California and Nevada deserts – the Pacific Coast Borax Company sold a brand called 20 Mule Team Borax, named after the method they used to transport marketable quantities from the desert.

In 1732, the French chemist, Claude François Geoffroy (also known as Geoffroy the Younger), noticed that a peculiar green flame was produced if he treated borax with sulphuric acid to produce boric acid, then added alcohol to set this on fire. This test showed that boron was present, and it came to be used as the standard way of identifying the presence of borax – it was crucial in the discovery of the Death Valley deposits.

In 1808, the French chemists Joseph Louis Gay-Lussac and Louis-Jacques Thénard, and Sir Humphry Davy in England, managed separately to extract boron after heating borax with potassium metal. This wasn't pure boron – that would only be isolated in 1909 by Ezekiel Weintraub in the USA – but it proved to be a brown amorphous solid with many useful chemical properties.

Life on Earth

While boron is present in some of the Earth's oldest rocks, it is only found in its pure metalloid form in meteorites. Nonetheless, it plays a crucial role in the biology of our planet – it stabilizes ribose, which is theorized to have played a key role in the development of DNA. And plants simply don't grow unless there are trace elements of boron in the soil – it is indispensable in the formation of plant stem cells.

Carbon

Category: non-metal

Atomic number: 6

Colour: clear (diamond), black (graphite)

Melting point: n/a (turns to vapour before it melts – a process of sublimation)

Sublimation point: 3,642°C (6,588°F)

First identified: *circa* 3750 BC

All life on Earth is carbon-based. Indeed, it is not known for certain that any other kind of life is even feasible. Carbon is a tetravalent atom, meaning it can bond to four different atoms at once, and this allows it to form as many as 20 million different compounds, and to form chains of different lengths.

Until the early nineteenth century it was thought that living matter as well as chemicals such as proteins or carbohydrates contained a 'spark of life' that made them completely different to inorganic matter. Then, in 1828, it was discovered that urea crystals, which are found in animal urine, could be synthesized in a laboratory, proving that there was no essential difference between organic and inorganic matter.

In the carbon cycle, photosynthesis by plants and plankton allows them to derive carbon from carbon dioxide, releasing oxygen as a by-product. At the same time, hydrogen combines with carbon to form carbohydrates.

These combine with nitrogen, phosphorus and other elements to form the molecules required for life, including bases and sugars for DNA and amino acids. Species such as humans, that don't photosynthesize, have to consume other plants or animals to get the carbon they need for their own cellular structures. And carbon returns to the start of the cycle either through being exhaled as carbon dioxide or through the decay of living matter after the cells die.

So, carbon is extremely important, even before you consider the many physical uses we have for its different forms. One of the most fascinating things about pure carbon is the huge difference between its allotropes. It occurs naturally in three allotropes, all of which have been known about since the time of the ancient Egyptians sixty centuries ago: diamonds, anthracite (a type of coal) and graphite. The only fundamental difference between these is in the atomic structure. However, diamonds are transparent and extremely hard, whereas graphite is black and soft: so how can they possibly be the same substance?

The answer is that many solids contain a 'crystal lattice' in their structure – the atoms are arrayed in a repetitive three-dimensional structure defined by the way the bonds hold them together. Diamonds have their atoms arrayed in a tight arrangement of three-dimensional tetrahedrons (pyramids with four triangular faces). By contrast, graphite's atoms are bonded equally tightly, but in two-dimensional layers that are only weakly connected to layers above and below them, which is why it seems soft. These atomic differences account for the very different appearances of the allotropes.

The connection between diamonds and other types of carbon was not understood for many millennia. In the seventeenth century, two Florentine scientists (Giuseppe Averani and Cipriano Targioni) discovered that it was possible to destroy a diamond by using a large magnifying glass to focus the heat of the sun on it. In 1796, Smithson Tennant, an English chemist, astonished the world by proving that diamond was merely a different form of carbon – he did this by demonstrating that when a diamond burned, carbon dioxide was the only product.

Carbon combines with hydrogen in strong, bonded chains to form hydrocarbons, which are extracted from the earth as fossil fuels, and also form the basis of plastics, polymers, and many fibres, solvents and paints. Global warming has been caused by the increased release of carbon dioxide from the burning of fossil fuels, and this will continue to be a problem until alternative, sustainable energy sources are available on a much larger scale.

Carbon is also crucial in many manufacturing processes – charcoal or coke is used to turn iron into steel; graphite is used in pencils (though wrongly called 'lead'), in electric motor brushes and furnace linings; diamond is used for cutting through rocks and drilling; carbon fibre is a particularly strong, light material used in fishing rods and tennis rackets as well as in moving parts in aeroplanes and rockets.

In recent years, scientists have even discovered ways to form carbon into exotic new allotropes with astonishing properties: fullerenes, which were identified in 1985, are hollow cages made of carbon atoms: the 'Buckyball', a type of fullerence, is shaped like a hollow ball of sixty carbon atoms; and nanotubes, which were discovered in 1991, are thin tubes only a nanometer (0.000001 mm) in diameter, formed from curled-up sheets of carbon atoms.

Perhaps the most amazing allotrope is graphene, which has been widely hailed as a miracle material of the future. While graphite is soft at a macro-scale, the individual sheets that make it up are incredibly hard. Scientists had theorized since the 1960s that it might therefore be possible to create a carbon material that was effectively two-dimensional and very light but extremely strong and flexible. In 2004, this extraordinary material, known as graphene, was finally produced – its possible uses include electric circuits, highly efficient solar cells, 'intelligent shoes', lightweight aeroplanes and even a new kind of neural implant that functions as a brain-to-computer interface.

The story of carbon is ongoing and about to become even more astonishing!

Nitrogen

Category: non-metal

Atomic number: 7

Colour: none

Melting point: −210°C (−346°F)

Boiling point: −196°C (−320°F)

First identified: 1772

N itrogen is a gas that makes up 78 per cent of the atmosphere of the Earth, and that we use in a huge variety of ways, from airbags to spray-on cream to the 'laughing gas' used in hospitals. However, we have only known of its independent existence for about 250 years.

Scientists in the eighteenth century became fascinated by the make-up of the air we breathe. The Scottish chemist Joseph Black had isolated carbon dioxide in the 1750s. He called it 'fixed air', as it could be released from minerals such as limestone when treated with acid. However, it was also known as 'mephitic air', meaning 'poisonous', because it seemed to suffocate animals that were immersed in it. If the oxygen in an enclosed space is all burned up, the remaining air has a similar effect, but in this case it is nitrogen (along with the other non-oxygen gases, including carbon dioxide, in the atmosphere) that is responsible.

Henry Cavendish was investigating mephitic air when he first discovered nitrogen, but the discovery is generally credited to Black's student, Daniel Rutherford, who carried out similar experiments and published his findings in 1772.

Cavendish, a methodical, slightly obsessive man, carried out numerous repetitions of an experiment in which he separated out the components of air. First, he passed air over heated charcoal, which converted the oxygen to carbon dioxide. Then he dissolved the carbon dioxide using an alkali solution. This left behind a separate gas, whose main component would later be named nitrogen ('nitre-forming') because it formed potassium nitrate (also known as saltpetre or nitre), a key component in early gunpowder.

Nitrogen played an important role in the development of explosives. Nitroglycerine, a liquid that will explode on impact, is formed when glycerine reacts with nitric acid. Alfred Nobel invented a much safer explosive, dynamite, when he found a way to absorb nitroglycerine into the soft rock 'kieselguhr' (also called diatomaceous earth).

The explosive nature of nitrogen is exploited when sodium azide is used in car airbags. A compound of sodium and nitrogen, it can be triggered by a spark to explode and decompose into nitrogen gas and sodium metal – the nitrogen rapidly inflates the bag.

Nitrogen can also come in handy if you want to freeze a piece of fruit. Liquid nitrogen is used in the flash-freezing process: you can watch a banana that has been frozen in nitrogen being smashed into tiny pieces with a hammer in various online videos. And nitrogen can be used to preserve fruit, too; if the fruit is stored in a non-refrigerated, sealed box of nitrogen, the types of decay that are caused by reactions with oxygen are prevented, and it will keep for up to two years.

Nitrogen is also used in the widgets that cause beer cans

to froth – a ball containing nitrogen, with a small hole in it, is left in the can, then a dash of liquid nitrogen is added to the beer in the compression process. This expands when the can is sealed, so that the nitrogen is compressed into the ball – when the can is opened, the pressure is released, triggering a burst of gas through the beer. Spray-on cream also relies on compressed nitrogen: in this case, nitrous oxide ('laughing gas') is absorbed into the cream, which is compressed in the can. When the compression is temporarily released, the pressure forces the cream out of the can.

Green plants and algae absorb nitrates, which help to form DNA and the amino acids that help in the creation of protein; thus nitrogen is also a key element for our living organisms. Animals consume nitrogen through their diet, and the nitrogen is eventually released back into the atmosphere. Microbes and bacteria in the soil then convert nitrogen back into nitrates (and this process can be encouraged by the addition of chemical fertilizers made from ammonia, which is a compound of nitrogen and hydrogen).

Oxygen

Category: non-metal
Atomic number: 8
Colour: none
Melting point: −219°C (−362°F)

Boiling point: −183°C (−297°F)
First identified: 1770s

Together with carbon, oxygen is one of the most crucial components of life on Earth. We breathe in, absorb and exhale carbon dioxide. Our brain, DNA and cells, and pretty much every molecule in our body, rely on oxygen, and it makes up about 60 per cent of our body mass, mostly in the form of water, or H_2O.

However, in spite of oxygen being the third most abundant element in the universe, it was something of an accident that our planet ended up with such a profuse supply. Before any larger animals existed, plants and cyanobacteria took their energy from the sun, absorbing carbon dioxide and exhaling oxygen. As oxygen is a highly reactive element, a large part of that oxygen reacted with other elements to form compounds. For example, did you know that 46 per cent of the mass of the rocks on the planet is oxygen – humble sand is actually silicon dioxide, many metals that we extract are taken from oxides (for instance, iron often comes from hematite, aluminium from bauxite), and carbonates such as limestone also contain oxygen.

In addition, the excreted oxygen entered the atmosphere,

of which it gradually reached a level of about 21 per cent, effectively terraforming planet Earth. It was dissolved oxygen in the water that allowed species to develop there, and over time life also evolved on land.

In the fifteenth century, Leonardo da Vinci had noticed that a candle wouldn't burn without air and speculated that it contained something life-giving. The discovery of oxygen was made independently by three chemists in the 1770s. In 1774, Joseph Priestley collected oxygen by focusing sunlight on mercuric oxide, and observed that the resulting gas made a candle burn more brightly (like his colleague Henry Cavendish, he wrongly identified this as 'dephlogisticated air'). In 1777, the Swedish scientist Carl Wilhelm Scheele published an account of discovering oxygen, which he had actually experimented with in 1771. And Antoine Lavoisier also identified oxygen – in fact, he recognized that this was indeed a new element, rather than air without phlogiston in it, although his name for it, 'oxy-gène', meaning 'acid-forming', was derived from the incorrect assumption that this gas would be present in all acids. In spite of that, the name stuck.

One of oxygen's less appealing qualities is that, because it is so reactive and helps to support many microorganisms, it causes many forms of decay, including rotting food. For many years, scientists have battled to come up with ever more inventive ways of keeping it away from food – fruit can be stored in nitrogen, buried, kept in tins or vacuum packs; it can be frozen, dried, cured or bottled to prevent oxygen doing its dirty work.

Relight My Fire

One of the classic chemistry experiments is a test for pure oxygen. First, you place some pure oxygen (or air with a high density of oxygen) in a flask. Then you light a wooden splint before shaking it to extinguish the flame – at this point it will be glowing slightly and will glow a bit more orange if you blow on it, but it will not relight. However, if you put it momentarily into the flask of oxygen, it will immediately relight, showing how reactive oxygen is and how easily it feeds flames.

At higher altitudes, the concentration of oxygen in the atmosphere is lower, which is why we find it harder to breathe there. And remember that oxygen doesn't just come in the dioxygen molecules we breathe (two oxygen atoms bonded together). It can also come as trioxygen, better known as ozone or O_3, which has three atoms. In the stratosphere, oxygen particles are constantly bombarded by UV radiation and split into individual atoms – these recombine with dioxygen molecules to form ozone molecules, which in their turn are hit by UV and split apart again. This is a continuous churning process and one that is crucial in protecting us from the sun's dangerous UV rays. When we are lucky enough to be able to watch the aurora or the Northern or Southern Lights, the beautiful swirling patterns are the result of the solar wind colliding with oxygen molecules way up in the atmosphere.

Fluorine

Category: halogen
Atomic number: 9
Colour: pale yellow
Melting point: −220°C (−363°F)

Boiling point: −188°C (−307°F)
First isolated: 1886

Fluorine is part of group 17 of the periodic table, the halogens, which also include chlorine, bromine, iodine and astatine. Halogen means 'salt-producing' – this is because halogens react with metals to give a range of salts, including calcium fluoride, sodium chloride (table salt) and silver bromide. They also share the properties of being highly reactive and potentially lethal. Fluorine is particularly dangerous in its pure form – if you breathe a concentration of air containing just 0.1 per cent fluorine it will be fatal within minutes, and if a stream of the gas is aimed at solids such as bricks or glass, they burst instantaneously into flames.

There are, however, safer compounds that contain fluorine. The mineral fluorspar (calcium fluoride) was in use by the 1520s as a 'flux' in furnaces – the way it melted and flowed when heated made the metal easier to work with. Alchemists of the period knew that fluorspar and fluorides in general contained an unknown substance, but they were unable to isolate it. (Or, if anyone did manage to, they probably died in the process and were unable to tell the tale!)

In 1860, the English scientist George Gore came close

to isolating the gas: he ran a current of electricity through hydrofluoric acid and may have safely produced a quantity of fluorine, but he couldn't prove it. It was not until 1886 that the French chemist Henri Moissan succeeded in using electrolysis to isolate fluorine (without dying in the process), a feat for which he was eventually rewarded with a Nobel Prize.

We do encounter fluorine on a regular basis in the stable form of fluorides – these substances are essential for humans and are added to water in many areas, following research that discovered that places with natural levels of fluoride in the water have lower levels of tooth decay. This practice is controversial, but fluoride is also in regular use in toothpaste – when applied to teeth, it forms tiny crystals that help make them more resistant to acids and to tooth decay.

Another common compound is polytetrafluoroethylene. This is a daunting name for the substance that you probably know better by the trademarked name Teflon. It was discovered in 1938 by Roy Plunkett in the DuPont laboratories, where he was researching new types of refrigerant (cooling gas). After storing the gas polytetrafluoroethylene in cylinders, he found that it left behind a residue of white powder.

The substance turned out to be a plastic that was heat-resistant, chemically inert and extraordinarily flexible at very low temperatures – this quality led to it being used in space exploration, while the fact that nothing would stick to it resulted in its widespread use in non-stick pots and pans. Teflon is also used in 'breathable clothing', which keeps the rain out while allowing water vapour to escape, making it ideal for anyone who exercises or works in rainy conditions.

Neon

10	
Ne	
20.1797	
Neon	

Category: noble gas
Atomic number: 10
Colour: colourless
Melting point: −249°C (−415°F)

Boiling point: −246°C (−411°F)
First identified: 1898

N eon is a great example of the way that Mendeleev's periodic table inspired chemists to search for elements they might not otherwise have been alerted to. Sir William Ramsay had already discovered other members of the group of noble gases (also known as the inert gases, due to their lack of reactivity), including helium, argon and krypton (see pages 10, 51 and 84), but the periodic table predicted another member in the spot vertically between helium and argon.

Working with his colleague Morris Travers at University College in London, Ramsay continued to search for the missing element. Having previously isolated argon, they now took a lump of solid argon, and surrounded this with liquid air – the argon evaporated slowly in low pressure conditions, and they were able to collect the first gas to evaporate. When they tested this in an atomic spectrometer, the heated gas gave off an extraordinary glow: Travers wrote, 'the blaze of crimson light from the tube told its own story and was a sight to dwell upon and never forget'. (A simpler method of fractional distillation is now used to extract neon from air.)

The Neon Spectrum

Neon only gives off that one vivid red shade of light. So why do we think of neon lighting as being multicoloured? The answer is that, because neon came first, it gave its name to a style of light – other colours are produced using different gases, coloured glass, or fluorescent powders baked onto the inside of the glass tubes. For instance, helium or sodium produce an orange light, argon a lavender light, krypton a blueish white or yellowish green, and for blue you can use xenon or mercury vapour.

Ramsay's thirteen-year-old son suggested calling the new gas 'novum' after the Latin for 'new': Ramsay adapted the idea, using the Greek-derived 'neon' instead. While it was a remarkable discovery, neon was initially a pretty boring element, since it is the least reactive of all: in fact, there are no elements that it will react with.

However, that brilliant red light fired up the imagination of the French chemist and inventor Georges Claude, who used an electric discharge in a closed glass tube containing neon to make an entirely new kind of light. His neon lamps were first displayed as a curiosity at a 1910 Paris exhibition. However, it took him over a decade to find a commercial use for the device (because people simply didn't want red light in their homes or streets). Once he worked out a way to use bent tubes to make letters that glowed, his company Claude Neon found success, in America in particular. The first neon signs were sold to a Los Angeles car dealership, where passers-by would stop to gawk at this amazing newfangled piece of advertising.

Sodium

Category: alkali metal
Atomic number: 11
Colour: silvery white
Melting point: 98°C (208°F)

Boiling point: 883°C (1,621°F)
First identified: 1807

At least two important sodium compounds have been used since the earliest civilizations. The ancient Egyptians harvested sodium carbonate ('soda') from dried-out flood plains near the River Nile – the crystals, which were used as a cleaning agent, are also mentioned in the Bible. And salt (sodium chloride), which was harvested from salt flats (or underground deposits), has always been an important part of our diet, whether added to meals or absorbed from animal-based food.

We have about 100 grams of sodium in our body – it is an electrolyte, like potassium and calcium, and thus is crucial in regulating inflows to and outflows from cells. It helps cells to transmit nerve signals, and to regulate the water levels in our body. However, too much sodium can dangerously raise our blood pressure, which is why high-blood-pressure patients are advised to reduce their salt intake.

Salt taxes have caused unrest throughout history – for instance, they were a contributory factor to the French Revolution. And, when the British Empire imposed a salt tax on India, where the largest part of the population was vegetarian, Mahatma Gandhi's salt march in protest became a key moment in the Indian independence movement.

Chemists in the Kitchen

Salt and soda have been with us for millennia, and caustic soda was first prepared by thirteenth-century soapmakers, but baking soda (sodium bicarbonate) is a relatively recent innovation. In 1843, Alfred Bird, a British chemist, created a batch to help his wife, who had a yeast allergy. It works by releasing carbon dioxide bubbles into the batter or dough through an acid-base reaction.

In spite of sodium being the sixth most common element on the planet and one whose compounds were extensively used by humans, it was only in the nineteenth century that its true nature was understood. It is extremely reactive, so it is never found naturally in its pure form (it tarnishes immediately if exposed to the air and can only be preserved beneath particular oils).

The first pure sodium was extracted by Sir Humphry Davy at London's Royal Institution – he ran an electric current through caustic soda (sodium hydroxide) and produced small globules of the metal. (Today it is generally produced by electrolysis of dry molten sodium chloride.)

It has many practical uses, including as a coolant in nuclear reactors, for de-icing roads in the winter (in the form of salt), clearing drains (in the form of caustic soda), and acting as a reagent (a substance that triggers reactions) in the biochemicals industry. However, most chemists know that the best fun you can have with a freshly sliced lump of sodium is to drop it into water and watch the resulting reaction (it bursts into flames before exploding) from a safe distance – the smartest way to see this is to watch a video on the internet, rather than trying it at home!

Magnesium

Category: alkaline earth metal

Atomic number: 12

Colour: silvery white

Melting point: 650°C (1,202°F)

Boiling point: 1,090°C (1,994°F)

First identified: 1755

Magnesium is the lightest metal that we can easily and safely use – lithium and sodium are both highly reactive, while beryllium is too toxic to be used without extreme precautions. It burns extremely brightly when ignited in air; a common science experiment in schools is to demonstrate this in the classroom, by lighting a ribbon of the metal.

Another element that is crucial for living organisms, magnesium plays an especially important role in photosynthesis as part of chlorophyll, the green plant pigment that does this job. You can spot a magnesium deficiency when plant leaves turn yellowy brown or develop unexpected dark red spots; the solution may be a leaf spray or the addition of calcium-magnesium carbonate to the soil.

Medicinal Magnesium

Epsom salts, which have been used since the seventeenth century to treat constipation, were discovered by a farmer who investigated why his cows (in Epsom in England) were ignoring a particular puddle during a drought; the puddle turned out to contain bitter-tasting crystals of magnesium sulphate, and the discovery inspired the medicine. Some readers will also be familiar with Milk of Magnesia, a suspension of magnesium oxide in water, used to treat indigestion and as a laxative.

We ingest magnesium from plants or other animals – especially from bran, chocolate, Brazil nuts, soya beans and almonds: it supports a range of bodily functions, including nerve and muscle functions, the regulation of blood sugar, and protein synthesis in the body. Some gastrointestinal diseases lead to magnesium deficiency, which can cause lethargy, depression and other more severe symptoms. It is also possible that a deficiency plays a role in Chronic Fatigue Syndrome (CFS)/myalgic encephalomyelitis (ME).

Joseph Black of Edinburgh recognized magnesium must be an element in 1755, following a careful comparison of magnesia (magnesium oxide) and lime (calcium oxide) extracted from the carbonate rocks, magnesite and limestone respectively. Humphry Davy isolated a tiny, pure amount of the metal in 1808 through the electrolysis of magnesium oxide.

Historically, magnesium was used to make meerschaum pipes (in the form of magnesium silicate), and its fierce flames were exploited in early flashlight bulbs and in the Second World War's terrifying magnesium bombs, which could cause huge conflagrations and firestorms – it is hard to ignite solid magnesium, so these bombs required a thermite reaction to ignite them. Happily, there are some more positive uses. The metal can safely be used, especially in alloys with aluminium and other light metals, to reduce the weight of metal components in cars and aeroplanes. It is also being used to create lightweight mobile phones and laptop computers.

Aluminium

Category: post-transition metal

Atomic number: 13

Colour: silvery grey

Melting point: 660°C (1,220°F)

Boiling point: 2,519°C (4,566°F)

First identified: third century or 1827

Aluminium is a hugely useful element that is used in everything from cans, foil, kitchen equipment and household items to aeroplanes, cars and power cables. It is a light metal that is soft and malleable, while being non-toxic, non-magnetic and an electrical conductor. Crucially, while iron rusts when it oxidizes, aluminium forms an extremely thin but tough layer of aluminium oxide, which only strengthens the metal. It is also the most abundant metal (by mass) in the Earth's crust, so is used extensively, in its pure form and in a variety of alloys; for instance, with magnesium, silicon, manganese and copper. These alloys are often used in aeroplanes, bicycles and cars that need to be lightweight.

Aluminium may have been refined as long ago as the third century, and the tomb of the Chinese military leader Chou-Chu contained a metal ornament that was 85 per cent aluminium. However, if the Chinese did have a method for partially refining the metal, it was lost for centuries. Eighteenth-century chemists had worked out that aluminium oxide must contain a metal, but it wasn't until 1827 that the German chemist Friedrich Wöhler perfected a method previously attempted by his Danish counterpart, Hans Christian Ørsted: aluminium chloride was heated with potassium, and this eventually produced pure aluminium.

Humphry Davy had come close to producing it twenty years later and had named the metal he was trying to refine 'aluminum' (with only one 'i'), from one of its compounds, the bitter salt alum. This led on to the current difference between American English and British English – over the years, the International Union of Pure and Applied Chemistry (or IUPAC) ruled that, as a metal, it should take the '-ium' ending, but the American Chemical Society chose to return to the original spelling, and now both countries think the other one is spelling it and pronouncing it 'wrongly'.

The Hall–Héroult Process

A cheap production method for aluminium was discovered independently by Charles Martin Hall, a twenty-two-year-old American amateur who carried out his ingenious experiments with his sister in a woodshed, and, on the other side of the Atlantic, Paul-Louis-Toussaint Héroult, a French chemist of the same age. The method, which was named jointly after them, was to dissolve aluminium oxide in a vat of molten sodium hexafluoroaluminate (better known as 'cryolite'), then separate the aluminium and oxygen with an electrical current. This is still the method used today for large-scale production.

Aluminium is now relatively easy to produce, but before modern methods were developed it was regarded as a luxury metal – in the 1860s, the court of Napoleon III of France is said to have served visiting kings and queens food on aluminium plates, while lesser nobles were only given gold plates to eat off.

Silicon

Category: metalloid
Atomic number: 14
Colour: metallic, slightly blue

Melting point: 1,414°C (2,577°F)
Boiling point: 3,265°C (5,909°F)
First identified: 1824

I f your first thought on seeing the name silicon is the miniaturized chips that computers use, you might be surprised that, by mass, silicon makes up 28 per cent of the Earth's crust (where it's the second most abundant element after oxygen). In nature, it is only found in compounds, so you would be more familiar with its various oxides – which include flint, sand, rock crystal, quartz, agate, amethyst and opal – and silicates, which include granite, asbestos, feldspar, mica and clay. The silicon in these compounds was originally formed by nuclear fusion inside dying stars before being ejected when the star collapsed into a supernova.

These compounds have all been used extensively throughout history. Some of humanity's earliest weapons were made from flint. Granite and other rocks used in construction are complex silicates. Sand (silicon dioxide) and clay (aluminium silicate) are key ingredients of concrete, cement, ceramics and enamels. Opals, quartz and amethysts were all valued by ancient civilizations. Glass, in the form of obsidian, occurs naturally in some places. By the second century BC, humanity had learned to manufacture it, after seeing how small droplets of glass were formed as a by-product of metalwork, when sand was melted only to solidify in a different form. And asbestos, a group of naturally occurring silicates, has been used for its fire-resistant qualities for millennia (although we use it increasingly cautiously nowadays, due to its carcinogenic nature).

Maybe it was the sheer variety of silicon's forms that led to it being mostly ignored by chemists until the nineteenth century. In 1824, the first relatively pure silicon powder was extracted from potassium fluorosilicate by the Swedish chemist Jöns Jacob Berzelius, but it wasn't until 1854 that the French chemist Henri Deville produced crystalline silicon.

Since then, silicon has become ever more useful – it is, for instance, used in alloys with aluminium and iron in transformer plates, engine blocks and machine tools. Mixed with carbon it forms silicon carbide, a strong abrasive. Together with oxygen it forms a polymer called silicone, which is a bit like rubber and can be used to seal bathrooms as well as, more controversially, in breast implants.

The name of Silicon Valley, the heartland of the digital world, is testament to the huge importance of silicon chips. These rely on silicon's status as a 'semiconductor', meaning that it will conduct electricity under certain circumstances, but not under others. The material used in chips is actually 'doped' silicon (meaning slightly adulterated with other elements to make it function like a kind of miniaturized transistor).

Science-fiction writers (and some scientists) have suggested there could be alien lifeforms based on silicon rather than carbon – the two are neighbours in the periodic table and, as with carbon, a silicon atom can bond with up to four other atoms at once. However, silicon-based life would probably require a very different kind of planet, with ultra-low temperatures and abundant ammonia.

Nonetheless, silicon does play some intriguing roles in terrestrial lifeforms – we don't really understand the role that phytoliths (tiny pieces of silica that form within plants) play, but they don't rot, so they survive in fossils, which makes them extremely useful to scientists. When you are stung by a nettle, it is through tiny silicate shards that are on the surface of the plant; and complex silicate structures are found inside one of the smallest photosynthesizing species, diatoms – tiny algae that produce huge quantities of oxygen on our planet.

So, who knows, maybe silicon-based alien life is not as far-fetched as it sounds!

Phosphorus

15

P

30.973761998
Phosphorus

Category: non-metal

Atomic number: 15

Colour: white, red, violet or black

Melting point (white): 44°C (111°F)

Sublimation point (red): 416–590°C (781–1,094°F)

Boiling point (white): 280°C (537°F)

First identified: 1669

Phosphorus was the thirteenth element to be discovered. It has sometimes been called the devil's element, due to the unlucky nature of the number 13 and to some of its other unpleasant properties. It was discovered in 1669 by the German alchemist Hennig Brand – while searching for the philosopher's stone, he boiled up and evaporated a huge vat of urine that had been left to ripen for days (which must have created a pretty vile smell). When he reheated the residue, it emitted glowing phosphorous vapour, which he condensed. It was Lavoisier, a century later, who recognized and named the element phosphorus, which means 'light bearer' in Ancient Greek. An easier production method used was to dissolve animal bones in sulphuric acid to form phosphoric acid, which is then heated with charcoal to produce white phosphorus.

Different allotropes and forms of phosphorus come in different colours – the most common ones are white phosphorus – which is poisonous, dangerously flammable in air, glows in the dark and can cause nasty burns on contact with skin – and red phosphorus – a safer type that you find in the material on the side of matchboxes. White phosphorus was used in matches such as those manufactured in Stockton-on-Tees in the UK from 1827 onwards; however, many of the girls who worked in the factory producing these were afflicted by 'phossy jaw', a nasty disease that caused the jawbone to waste away, and so this use was banned early in the twentieth century.

Phosphorus has been used in some deadly weapons, from tracer bullets, fire bombs and smoke grenades to the phosphorus bombs that caused terrible firestorms in Hamburg in 1943. It has also been used in nerve gases such as sarin, which killed or damaged many people in the Iraq–Iran war in the 1980s (as well as killing twelve and harming many other people in the Tokyo subway attack in 1995).

Happily, you don't find phosphorus in nature – only phosphates, which are crucial to life in a variety of ways and found in DNA, tooth enamel and bones: we consume them in foods such as tuna, eggs and cheese. Phosphates are also used as fertilizers, which have enabled humanity to hugely increase agricultural production in recent centuries. But we may be heading for serious problems in the phosphorus cycle. On the one hand, using too many phosphates in fertilizers or detergents has led to pollution of rivers and lakes, which encourages algae to grow, harming water-dwelling organisms that rely on photosynthesis and, as a result, other species that need the oxygen they produce. At the other end of the scale, we may be within a few centuries of running out of sources of phosphate. Guano, manure and human excreta were sources of phosphates in the past, but the only economically viable source now is phosphate rock from a limited number of places.

Enemy of the People

Antoine Lavoisier was a member of the French establishment, working for the customs and tax organization, the *Ferme générale*. His political connections helped fund his brilliant research but contributed to his downfall in the end. In 1794, after the French Revolution, he was charged with tax fraud during the Reign of Terror and went to the guillotine.

At the same time as we have been pumping out too much phosphorus into the environment (in forms that aren't easy to reclaim as usable phosphates), we have been removing the phosphates that support current levels of food production at an accelerating rate. Many scientists expect this to be one of the major environmental crises of the next one hundred years, even if the issue is currently below most people's radars.

Sulphur

16
S
32.06
Sulfur

Category: non-metal
Atomic number: 16
Colour: yellow
Melting point: 115°C
(239°F)

Boiling point: 445°C
(833°F)
First identified:
prehistory

Another element associated with the devil, sulphur (also known as 'brimstone') gets fifteen mentions in the Bible and is credited with the destruction of Sodom and Gomorrah, although its bad reputation may be inspired by the terrible smell of some of its compounds rather than anything essentially devilish about it. The element is found naturally, often as bright yellow crystals attached to rocks in volcanic areas. It was used historically to bleach cloth and preserve wine by burning it to obtain sulphur dioxide, which was passed through the clothes, or grape juice. When unpurified fossil fuels are burned today, the resulting sulphur dioxide in the atmosphere can cause acid rain or exacerbate smoggy conditions in cities.

Alchemists believed that all metals contained sulphur, mercury and salt, so they carried out many weird and wonderful experiments with it. The evil smell comes when sulphur is in the form of hydrogen sulphide or any one of the sulphide compounds, called 'mercaptans' – when an egg goes off, globulin (a protein that eggs are high in) decays and produces hydrogen sulphide, which has a truly noxious smell. Skunks defend themselves by excreting 'butyl seleno-mercaptan', while other milder mercaptans are added in small quantities to odourless natural gas so that it has an unpleasant smell (and an inbuilt warning device for gas leaks).

Sulphur to the Rescue

While there is much debate about the detail, some scientists believe that sulphur will play a crucial role in mitigating global warming. The compound dimethylsulphide is indirectly produced by plankton in the oceans of the Earth – this eventually oxidizes to sulphuric oxide, which leads to sulphuric acid particles entering the atmosphere, where they help in the process of cloud formation. So higher temperatures may naturally lead to a feedback mechanism, which will have a cooling effect in response.

Sulphur does, however, have many positive roles – its compounds are used to vulcanize rubber, bleach paper, produce phosphates for fertilizers, and as preservatives and detergents. Calcium sulphate is a crucial ingredient in cement and plaster. Sulphuric acid is especially important to industry and is used, for instance, in the production of phosphates for fertilizers. Sulphates from the soil also enter into the ecosystem and are crucial for several types of amino acids and enzymes – there are currently about 150 grams of sulphur compounds in your body.

Chlorine

Category: halogen

Atomic number: 17

Colour: yellowy green

Melting point: −102°C (−151°F)

Boiling point: −34°C (−29°F)

First identified: 1774

You will be very familiar with chlorine in the everyday form of salt (or sodium chloride), which is a crucial part of our nutritional needs as part of the foods we eat. However, as with other elements, chlorine has more than one face: when the German army used chlorine gas as a weapon at Flanders in 1915, it resulted in about 5,000 deaths and many more suffered terribly from its effects.

Chlorine is not found in its pure form in nature – it was first isolated in 1774 by Swedish chemist Carl Wilhelm Scheele, who heated hydrochloric acid with manganese dioxide. The result was a yellowy-green gas with a foul, choking smell, which could be dissolved in water to give an acid solution. Scheele didn't believe that the gas he had produced was pure, but in 1807 Humphry Davy carried out further investigations and announced the existence of a new element (which was named after the Greek word for yellowy green).

One compound of chlorine is PVC (polyvinyl chloride), that versatile plastic used for everything from window frames to blood bags in hospitals. It is also used widely in the pharmaceutical industry to trigger chemical reactions. But chlorine is probably best known for its role as a disinfectant, as it kills bacteria; it is present in many household bleaches and is used to make tap water and swimming pools safe. The latter use originated in London, following a cholera outbreak, when the pioneering doctor John Snow realized that an infected well in Soho was the cause. He attempted to disinfect the pump from the well with chlorine. For the rest of the century there were various isolated cases of this kind of use of the chemical, before comprehensive chlorination of drinking water started to be adopted in Europe and America early in this century. Snow also experimented with the chlorine compound chloroform and used it while helping Queen Victoria to give birth.

Our attitudes to some chlorine products have changed significantly over time. Chloroform and the dry-cleaning solvent, carbon tetrachloride, were in common use in the past, but are now carefully controlled as they can damage the liver. At one stage, chlorofluorocarbons were also widely used, especially in aerosols – better known by the acronym CFCs; they were implicated in the ongoing destruction of the ozone layer. Since the 1980s there has been a huge global reduction in their use, which has happily helped the ozone layer to stabilize over recent years.

Argon

Category: noble gas
Atomic number: 18
Colour: colourless
Melting point: −189°C (−308°F)

Boiling point: −186°C (−302°F)
First identified: 1894

These days we are very conscious of the increasing amount of carbon dioxide in the atmosphere; we know all about the environmental problems that this will cause in the future. However, it is not as widely known that the air we breathe every day actually contains more argon (1 per cent) than carbon dioxide (0.4 per cent).

Our first glimpse of argon came in the 1760s during Henry Cavendish's investigations of the make-up of air. We've seen how Cavendish separated 'phlogisticated air' from 'dephlogisticated air'. Each time, Cavendish was puzzled to find that when he went on to extract nitrogen from the phlogisticated air, there was a stubborn residue of about 1 per cent of inert gas.

This was largely forgotten until 1894, when John Strutt (later known as Lord Rayleigh) and William Ramsay established that nitrogen extracted from air was always about 0.5 per cent more dense (and thus 'heavier') than nitrogen extracted from ammonia. They identified the heavier gas that remained after oxygen and nitrogen were removed from atmospheric air as a separate element – it was named after the Greek word *argos*, meaning 'lazy', because of its chemical inertness. This inertness is down to the fact that argon has a full outer shell of electrons and is thus one of the 'noble gases', which tend not to bond or easily react with other elements.

Discovery Delayed

When Strutt and Ramsay discovered argon in 1894, they didn't immediately broadcast this to the world, but not because there was any problem with their scientific results. The pair had realized that a major competition for chemical discoveries, being held the following year in America, came with the condition that the discovery had to be made after the start of 1895. Tempted by the money on offer, they sneakily held their discovery back and only announced it the following year. They duly collected $10,000, which would be worth well over $150,000 today.

Argon is an important ingredient in many industrial processes. For instance, in steel production it is combined with oxygen and blown through molten metal during 'decarburization': this prevents valuable elements in the steel, such as chromium, from oxidizing in large amounts. It is used in traditional incandescent lightbulbs, again because it doesn't react easily and prevents the filament from oxidizing at high temperatures. It is often used in double glazing to fill the gap between two panes of glass, because it is heavier than air and conducts less heat, so helps insulate the home.

More recently, blue argon lasers have been used in hospitals to destroy cancerous growths and to correct corneal defects in patients' eyes.

For a substance that we didn't even know about 125 years ago, argon has turned out to be a very useful gas indeed.

Potassium

Category: alkali metal
Atomic number: 19
Colour: silvery grey
Melting point: 63°C (146°F)

Boiling point: 759°C (1,398°F)
First identified: 1807

Many centuries ago, people discovered that potash was a useful fertilizer that could be prepared from a variety of plants. One eighteenth-century account of its production reads 'Potas or Pot-ashes, is brought yearly by the Merchant's Ships in great abundance from Coerland [now part of Latvia and Lithuania], Russia, and Poland. It is prepared there from the Wood of green Fir, Pine, Oak, and the like, of which they make large piles in proper Trenches, and burn them till they are reduced to Ashes.' The ashes were boiled in water, then the liquor at the top was poured into large copper pots and reboiled to reduce it to a salt. A similar account from the seventeenth century describes the same process being applied to a herb called 'Kali' (whose Latin name is *Salsola kali*, which is better known as saltwort).

Both methods produce a mix of potassium carbonate and sodium carbonate. (Potash can also be used to describe any of potassium chloride, potassium sulphate or potassium nitrate.) The kali method, which produces more of the sodium compound, is the origin of the word 'alkaline', via 'al-kali' where 'al' is the Arabic definite article.

This is relevant, because potassium was the first of the alkali metals to be isolated: Humphry Davy succeeded in doing this in 1807 by electrolysing molten potassium hydroxide. Reportedly, when he first saw 'the minute globules of potassium burst through the crust of potash, and take fire as they entered the atmosphere, he could not contain his joy'.

Minding your 'P's and 'Q's

The brilliant Swedish chemist Jöns Jacob Berzelius invented the system of chemical notation we use today, using abbreviations such as O for oxygen. The only difference is that Berzelius used a superscript (H^2O) for numbers of atoms rather than a subscript (H_2O). You may have noticed that some of the elements have abbreviations that seem to have nothing to do with the name – this is usually down to historic disagreements between various European chemists; for instance, sodium (Na) was known as natrium or natronium in Germanic languages. Davy named potassium after potash, but Berzelius preferred 'kalium' after the plant kali, thus the chemical symbol K.

Potassium is light enough to float on water – although if you try this, you'll find it explodes almost immediately as it is so reactive – in fact it will even burn a hole through ice. Its main industrial use today remains as a fertilizer, as plant cells need potassium (and it is a crucial part of our diet). It is also used in the manufacture of glass, liquid soap, pharmaceuticals and saline drips.

Calcium

Category: alkaline earth metal

Atomic number: 20

Colour: silvery grey

Melting point: 842°C (1,548°F)

Boiling point: 1,484°C (2,703°F)

First identified: 1808

O ver the years, advertising for milk has taught us how important calcium is for keeping our bones and teeth in good health. We also obtain calcium from foods such as cheese, spinach, almonds, fish, seeds and yogurt. So, it may not be immediately obvious that calcium is a metal.

It is actually the fifth most abundant metal in the Earth's crust, but you don't find it in its natural form – it reacts rapidly with air and forms various compounds, including limestone (calcium carbonate) and fluorite (calcium fluoride), and manmade compounds such as gypsum or plaster of Paris (calcium sulphate), lime or quicklime (calcium oxide). Chalk is a type of limestone, and stalactites and stalagmites are formed in caves when water containing dissolved calcium bicarbonate drips from the ceiling and the mineral content gradually precipitates back into limestone. When drinking water is described as 'hard' rather than 'soft', this indicates a higher mineral content, which is predominantly calcium compounds dissolved when the water flowed over limestone or other minerals. Hard water is responsible for the calcium deposits that clog up kettles and washing machines, but is relatively harmless (and apparently produces better-tasting beer than soft water!).

Bone-builder

In our bodies, calcium is continually at work in helping to renew our bones. This is a constant process, but it slows down in pregnant women and with age – a lack of sufficient calcium is one of the causes of osteoporosis, in which bones become decalcified. It is also crucial in the process that allows blood to clot when you are bleeding.

Gypsum and lime have been used since antiquity. Lime is an important ingredient in mortar or cement, and was used by the Romans and, before them, the ancient Egyptians – for instance, in the construction of the Great Pyramid of Giza. Gypsum was used to set bones many centuries ago, just as it is today.

Scandium

Category: transition metal

Atomic number: 21

Colour: silvery white

Melting point: 1,541°C (2,806°F)

Boiling point: 2,836°C (5,136°F)

First identified: 1879

endeleev left four notable gaps in his periodic table and speculated that an element would eventually be discovered to fill each of them. He gave the missing elements names based on their proximity to other known substances – in his original table, boron was at the top of group 3, and in the space for an element with an atomic number of 21 beneath that he added 'eka-boron' (by which he meant 'one place away from boron'). Boron was eventually moved to the top of group 13, but within a decade of the first publication of Mendeleev's table, the discovery of scandium, filling the 'eka-boron' space which was now top of group 3, helped to solidify his reputation and to focus worldwide attention on his work. (Gallium had already filled the first gap when it was discovered in 1875 – see page 74.)

The Swedish chemist Lars Fredrik Nilson managed to isolate a tiny sample of scandium oxide from the mineral euxenite in 1879, although it would take until 1937 for a larger sample of the (nearly) pure metal to be extracted from ores. It was named after Scandinavia, where it had been found. It is rare – global production is still only about ten tons a year.

This makes it considerably more valuable than gold, but it is mainly mined for industrial purposes rather than being traded as a luxury metal. It is a light element that produces excellent alloys, with aluminium in particular – these are used in lightweight sports equipment and aircraft. Scandium has also been used, in the form of scandium iodide, in high-performance floodlights.

While scandium is rare on Earth, it is more common elsewhere in the universe – both the sun and the moon have higher concentrations of the metal than our planet.

Titanium

Category: transition metal

Atomic number: 22

Colour: silver

Melting point: 1,668°C (3,034°F)

Boiling point: 3,287°C (5,949°F)

First identified: 1791

A lot of products sell better if they are a brilliant white – whether it be toothpaste, confectionery, paint or medicines. One of the best ways of achieving this is by using titania, or titanium oxide (also known as titanium white), a naturally occurring oxide of titanium.

Titanium oxide is also at work in one of the most useful inventions of the twenty-first century – for instance, used as a coating for self-cleaning glass in car wing-mirrors, where the water not only spreads out without misting, but even takes most of the dirt with it. The first widely available version of such glass was Pilkington's ActivT, which came on to the market in 2001.

Titanium is quite common – it is Earth's ninth most abundant element. However, it is not easy to extract – it reacts with nitrogen, which rules out a lot of possible methods. The current standard method is the Kroll process, which involves heating up titanium dioxide to about 1,000°C, then passing chlorine over it to form a different compound – this is then covered in argon and a reaction with magnesium at 850°C allows extraction of the pure metal. Though this is economically viable, it makes titanium a more expensive metal than abundant ones, such as iron.

Titanium is extremely useful because it is as strong as steel but less than half the weight, doesn't corrode in water and isn't subject to metal fatigue – like aluminium, it reacts with oxygen, but the result is a thin, oxide layer that protects the metal. As a result, it has many applications in transport, sports equipment and sea craft. It also bonds well with bone, which makes it ideal for hip replacements and tooth implants.

An oxide of titanium was first identified by a Cornish vicar named William Gregor in 1791 – he named the black sand 'menachanite' after the local parish of Manaccan. Just a few years later, the German chemist Martin Heinrich Klaproth realized that menachanite contained the same new element as rutile (a red ore), but it took until 1910 for chemists at General Electric in the USA to find a rudimentary method for extracting pure titanium metal.

Vanadium

23	
V	
50.9415	
Vanadium	

Category: transition metal

Atomic number: 23

Colour: greyish silver

Melting point: 1,910°C (3,470°F)

Boiling point: 3,407°C (6,165°F)

First identified: 1801

Vanadium is another metal frequently used in alloys. Eighty per cent of total production each year is added to steel, which becomes resistant to shocks and vibrations through the addition of less than 1 per cent of vanadium and a pinch of chromium. Alloys of the metal are also used in nuclear reactors, because of its low absorption of neutrons. And it has additional uses as a pigment for glass and ceramics, and in the manufacture of superconducting magnets.

The metal was first discovered in 1801 by Andrés Manuel del Rio, a Mexican professor, within a brown lead ore called vanadite. When he sent it off for further analysis, a French chemist dismissed it as being chromium (which it does resemble, due to the many colours of its salts). Then, in 1831, the Swedish chemist Nils Gabriel Sefström discovered that cast iron made from ore mined in southern Sweden also contained vanadium. The metal workers had wondered why there was so much variation in how hard their iron was: vanadium was the answer.

There were various attempts to isolate the metal over the years, and a few attempts to claim success: finally, in 1869, Sir Henry Roscoe produced a sample in Manchester, and demonstrated that the previous samples made had all actually been the compound vanadium nitride. The metal is generally produced by reducing vanadium oxide with calcium in a high-pressure environment.

Vanadium is an essential part of our nutritional needs, but only in tiny amounts – good sources are mushrooms, shellfish, spinach, whole grains, black pepper, dill seeds and parsley. It also seems that a healthy intake of vanadium is useful to diabetics, as it increases their sensitivity to insulin.

Chromium

Category: transition metal

Atomic number: 24

Colour: silver with a blue tinge

Melting point: 1,907°C (3,465°F)

Boiling point: 2,671°C (4,480°F)

First identified: 1798

Siberian red lead (or crocoite) is an orangey-red mineral that was discovered in the eighteenth century. In 1798, the French chemist Louis Vauquelin showed that it contained a previously unknown element, which he would name 'chromium' (from the Greek word for 'colour'), because of the beautiful and varied colours of its compounds.

Vauquelin did not expect his discovery to be of much use other than for decorative purposes, and to a degree he was right. Only a small proportion of the chromium that is produced is used in its natural form; chromium (or 'chrome') plating is used to give a shiny finish to steel (such as on some classic cars and bikes) and also to plastic household fittings. It is more commonly used in alloys or compounds. Steel and chromium are alloyed to form stainless steel, which develops a thin protective oxide layer rather than rusting as unalloyed steel would.

Chromium compounds create a remarkable variety of colours and shades for paint pigments: various types of chromium oxide, lead chromate, sodium chromate, chromium chloride and anhydrous chromium chloride produce shades of dark red, orange-red, bright yellow, light blue, and pale or dark green. Chrome yellow is a colour that was especially well known to generations of American children, as traditional school buses were painted with it so that they were easy to spot in gloomy conditions. (The traditional paint has since been replaced, as it contained lead and other toxic substances.)

Chromium also plays a rather beautiful part in the colouring of gemstones. Corundum and beryl are naturally colourless oxides, but if there is a tiny trace of chromium in the mix you get rubies and emeralds. The transformation of chrysoberyl is even more extraordinary: it is a colourless aluminate of beryllium but, if it contains a dash of chromium, it becomes alexandrite – a highly pleochroic gemstone (meaning that it absorbs different wavelengths depending on the direction of the light). The colour of a good-quality alexandrite specimen can change from orangey red to yellow to emerald green in different positions and lighting conditions.

Manganese

Category: transition metal

Atomic number: 25

Colour: silvery

Melting point: 1,246°C (2,275°F)

Boiling point: 2,061°C (3,742°F)

First identified: 1774

The famous cave paintings of Lascaux, in France, demonstrate one of humanity's first uses of a manganese compound: the black ore, manganese dioxide (or pyrolucite), makes a good black paint. Another compound, magnesium oxide, was used in Ancient Egypt to remove the green tint of glass (it was known as *sapo vitri*, meaning 'glass soap').

In its elemental form, manganese is a brittle, hard, silvery metal. About 1 per cent of it is used in the 'Bessemer process' method of steelmaking – it converts iron sulphide to manganese sulphide in the alloy – the latter has a much higher melting point. If you increase the proportion of manganese to about 13 per cent, you get 'manganese steel', which is extremely strong and is used in railway tracks, safes and the bars of prison cells. It is also alloyed with aluminium in drinks cans to lessen the risk of corrosion.

In the early eighteenth century it was assumed that manganese contained iron, but after a Berlin glassmaker demonstrated this must be untrue, various chemists attempted to isolate it before Johan Gottlieb Gahn's successful 1774 attempt in Sweden. (Gahn may actually have been beaten to the post by a few years by a Viennese student, who apparently produced the metal but failed to publicize his findings.)

Rolling on the Ocean Floor

Did you know that there are millions of nodules containing a high proportion of manganese littered on the ocean floor? In some places, they cover as much as two-thirds of the seabed: the actions of certain sea creatures seem to prevent them from being buried in sediment. There are various theories about the formation of these nodules, but scientists agree that they have taken millions of years to 'grow'. There has been a lot of interest in them from mining companies, since they could theoretically be harvested from the ocean, but this hasn't happened, due to the high costs of doing so, and increasing concern about the potential environmental impact.

The name can be confusing. Both magnesium and manganese were named after Magnesia in north Greece (where *lapis magnis* or magnetic lodestones were first discovered). At one point, magnesium was known as white magnesia, while manganese was called black magnesia. This confusion was not cleared up until after Gahn's breakthrough.

Manganese is the second most abundant transition metal after iron and appears in hundreds of minerals. It also plays a crucial role in photosynthesis and the development of certain enzymes: we consume trace amounts from foods like nuts, bran, wholegrain cereals, parsley and – most importantly from a British point of view – a nice cup of tea.

Iron

Category: transition metal

Atomic number: 26

Colour: silvery grey

Melting point: 1,538°C (2,800°F)

Boiling point: 2,861°C (5,182°F)

First identified: ancient civilizations

One of the most important metals in human history, iron is also the most plentiful element (by mass) on Earth, partly because the planet's core is largely made up of it. In the process of the Earth's formation, as a swirl of dust and gas was gradually compressed from a protoplanetary disc into a planet, the heaviest elements naturally gathered in the centre and iron ended up forming a solid, inner core and a molten liquid outer core. It is the iron core that gives the planet a magnetic field – there are magnetic poles at the north and south of the globe, and the field reaches out into space, deflecting potentially harmful solar winds and radiation.

The first iron artefacts date back to the ancient Egyptian period, but it was the Hittite civilization of Asia (in present-day Turkey) that discovered how to smelt the metal, which made it far easier to work, in about 1500 BC. The Hittites kept their discovery secret for centuries, but after their empire was invaded in 1200 BC, the ironmakers scattered, taking their skills with them and thus triggering the Iron Age.

Iron can be cast, welded or machined into a huge variety of forms – at various points throughout history, people developed the ability to create stronger, less brittle forms of iron by smelting it with carbon or other metals. The legendary metal, Damascus steel, was astonishingly hard, shatterproof and sharp, probably due to the fact that there was a proportion of vanadium in the ore used to make it (which came from the Indian subcontinent). It was the discovery of more efficient methods of production in the seventeenth century that triggered the huge surge of new technology, known as the Industrial Revolution, and this was developed further after the invention of the Bessemer process (a large-scale steel production method) in 1856. Since then it has been used to make everything from bridges to ships, skyscrapers to cars, tools to paper clips.

The biggest drawback to iron is how easily it rusts when it comes into contact with oxygen. This problem can be mitigated in a variety of ways, including coating the iron (or steel) with tin, or zinc (in which case it is 'galvanized'), or alloying it with nickel to make it resistant to corrosion.

Iron is another element that is crucial to life – it is present in our body in a variety of forms, notably in the haemoglobin that transports oxygen through the blood. A lack of iron in the diet leads to lower red blood cell production and anaemia, in which you become tired and breathless. The best sources of iron are red meat and liver, certain dried fruits, bread and eggs.

All elements heavier than iron, including gold and lead, were originally formed exclusively in supernovae rather than stars. Stars simply aren't hot enough to make the heavier elements.

Cobalt

Category: transition metal

Atomic number: 27

Colour: silvery blue

Melting point: 1,495°C (2,723°F)

Boiling point: 2,927°C (5,301°F)

First identified: 1735

Cobalt has been used for its colouring properties for between four and five millennia: the ancient Egyptians used it to create cobalt-blue paint and coloured necklaces, while the tomb of Tutankhamun (who was pharaoh from 1361 to 1352 BC) contained a deep-blue glass object coloured with cobalt minerals. It was also used to create pottery glazes. Cobalt chloride makes a blue or green colouring, or rose red when it is hydrated. And one of its quirkier applications is in the creation of invisible ink, which is made by dissolving the compound in water with glycerol – the resulting mixture is an ink that only appears once the paper is heated.

Cobalt doesn't appear naturally on its own, only in minerals and often in tandem with other transition metals, especially copper and nickel. It is mainly produced as a by-product of copper mining. It is also present in those weird manganese nodules that litter the ocean floor (see page 64). Its main biological significance is that it is part of vitamin B12, which we need in our bodies (we generally consume it in animal products or in fortified breakfast cereals).

The Blue Meanies

Cobalt was first isolated by the Swedish chemist Georg Brandt in 1735 – there was some controversy over whether he had produced merely a compound of iron, but in the end his discovery was accepted. He named it after the German word *kobold*, which means goblin. This is because German mineworkers in the previous century hated ores that contained the metal, which they sometimes mistook for silver – the high melting point made it impossible to work with, but it released toxic arsenic fumes when heated, so they regarded it as being an evil goblin's trick.

In the last century, cobalt has found some significant new applications. It is an ultra-strong, hard and magnetic metal, with an unusually high melting point. It is one of the three transition metals that are magnetic (along with iron and nickel). This makes it useful in alloys that need to be wear resistant, such as drills and saws. And because it retains its magnetism at high temperatures, it is often used in alloys for parts of high-speed motors.

Nickel

Category: transition metal

Atomic number: 28

Colour: silvery white

Melting point: 1,455°C (2,651°F)

Boiling point: 2,912°C (5,274°F)

First identified: 1751

A fter iron, nickel is the next most significant part of the Earth's core and continues to arrive on the planet in meteorites: two of the most significant deposits of the element come from areas near Ontario in Canada and Sudbury in England, which were the sites of past meteorite strikes.

In compounds, it plays a small role in our diet – in tiny quantities it is used to hydrogenate vegetable oil, and baked beans are weirdly rich in nickel. However, it is mainly used in alloys: the five-cent piece in the USA, known as a nickel, actually comprises 25 per cent of the metal and 75 per cent copper (nickel is also found in many other coins). In toasters and electric ovens you will find nichrome, an alloy mainly made of nickel and chromium, which doesn't corrode even when it is red-hot. Alloyed with steel and chrome, it creates stainless steel. With copper, it can be used in desalination plants. Morel is a compound of steel and nickel so extremely resistant to corrosion that it can even survive exposure to the destructive gas fluorine (see page 30). And superalloys made of aluminium and nickel, with a small amount of boron, are used in rocket turbines and aeroplanes, because they are very light but get tougher rather than weaker at high temperatures.

Nickel is named after St Nicholas's copper (or 'kupfernickel'), an ore found by German miners. The Swedish chemist Axel Fredrik Cronstedt managed to isolate nickel in 1751, although it would be several years before the scientific establishment believed that this truly was a new element rather than another alloy.

Copper

Category: transition metal

Atomic number: 29

Colour: reddish orange

Melting point: 1,085°C (1,985°F)

Boiling point: 2,562°C (4,644°F)

First identified: ancient civilizations

Copper is a native metal: this means it can be found in its pure or alloyed form in nature. For this reason, it was one of the first metals to be used by people – we know it was in use about 10,000 years ago (copper jewellery has been found in Iraq from the period), was being smelted from sulphide ores about 7,000 years ago and was then moulded into shapes about 6,000 years ago. Crucially, it was also the first metal to be deliberately alloyed with another metal – when combined with tin it forms bronze. This discovery marks the point at which metal tools started to replace stone ones, in the Bronze Age, which began in about 3500 BC. Along with silver and gold, it has frequently been used to make coins, although copper coins have tended to be the lower-value denominations.

The chemical symbol Cu derives from Roman times – it was known as *aes cyprius* (metal of Cyprus, location of the largest copper mines of the period). This was gradually corrupted to *cuprum* in Latin and copper in English.

Copper has an unusual reddish colour and is a hardy metal – when archaeologists excavated the Great Pyramid of Giza, they found copper tubes, part of a plumbing system, that were still in usable condition. It is also often used in wiring – it conducts heat and electricity well and can be stretched out relatively easily – and in building (especially piping and roofing), as well as in decorative art – statues and other artefacts made of it become covered in a green verdigris (patina) as the metal oxidizes. Compounds of copper, especially copper salts, also give a green or blue colour to minerals such as azurite or turquoise, and it has historically been used to produce paint pigments in these colours.

The human body requires trace amounts of copper. However, a more interesting fact is that, while most fish and mammals have some kind of haemoglobin, an iron complex, arthropods and molluscs need copper for the same purpose – their blood contains the copper complex hemocyanin instead of haemoglobin.

Zinc

Category: transition metal

Atomic number: 30

Colour: blueish white

Melting point: 420°C (788°F)

Boiling point: 907°C (1,665°F)

First identified: 1746

P inning down when an element was 'discovered' can be tricky – sometimes it is best to count the date when it was first refined by a chemist, while for other elements the appropriate date may be when it was first found or identified. We know that the Romans used zinc, and there is archaeological evidence that it was being refined in India from the twelfth century until the sixteenth century. Historians generally record the German chemist Andreas Marggraf as being the first to recognize it as a new element in 1746, but the Flemish metallurgist P. Moras de Respour had written about the extraction of the metal from zinc oxide in the previous century.

The most common use of zinc is for galvanizing steel to prevent corrosion – this process was invented by Luigi Galvani (who is also remembered for having used an electric current to make frogs' legs twitch). The usual method is hot-dip galvanization, in which iron or steel is briefly submerged in liquid zinc: this creates a thin layer, which protects the metal beneath.

The Rooftops of Paris

When Baron Haussmann renovated the city of Paris in the nineteenth century, for the rooftops he mainly used an alloy that was 80 per cent zinc. The beautiful silvery-grey roofs have become a signature image of the city, inspiring many artists and film-makers. They have recently been recognized as a 'priceless cultural asset' and are likely to be given Unesco World Heritage status. They have the advantage of being environmentally friendly – rainwater that runs off zinc does not leach up any of the metal (unlike lead and other heavy metals used for the same purpose).

Zinc can also be alloyed with copper to produce brass, which is widely used for everything from doorknobs to zips and, of course, the brass instruments in an orchestra. Its compounds include zinc sulphide, which is used for making paint and fluorescent lights, and zinc oxide, used in a wide variety of products, most notably as the main ingredient of calamine lotion.

Gallium

Category: post-transition metal	**Melting point:** 30°C (86°F)
Atomic number: 31	**Boiling point:** 2,229°C (4,044°F)
Colour: silvery white	**First identified:** 1875

Gallium was the first of Mendeleev's 'missing elements' to be discovered – he had predicted there would be an element to fill the missing space under aluminium and called it 'eka-aluminium' as a result. Within five years, the space had been filled: the French scientist Paul-Émile Lecoq de Boisbaudran was unaware of Mendeleev's prediction when he examined a piece of zincblende (or sphalerite) ore using spectroscopy and noticed two unusual violet lines in the spectrum it produced. He isolated this element and named it gallium (after the Latin name for France, although there is also a possible pun on his own name, Lecoq, as *gallus* is Latin for 'rooster'). Gallium is also found in a variety of other minerals, including bauxite, but is generally produced as a by-product in the making of various metals (for instance, when aluminium is refined from bauxite).

You can melt a piece of solid gallium in your hand, so low is its melting point – some scientists have taken advantage of this by making gallium spoons, which can be used as a practical joke (since they will melt when used to stir tea or coffee). It is often used in preference to the more toxic mercury in thermometers. And it has extremely useful semiconductor properties: in the form of gallium arsenide, it gives a faster performance than the traditional silicon semiconductor in a chip. It can be alloyed with most metals and is useful if a low melting point is required for the resulting alloy. It also has a few medical applications – the radioactive isotope gallium-67 can be used to diagnose and locate cancerous growths, while there is ongoing work on the use of a gallium compound in a new generation of anti-malaria medication.

Germanium

32
Ge
72.630
Germanium

Category: metalloid

Atomic number: 32

Colour: greyish white

Melting point: 938°C (1,720°F)

Boiling point: 2,883°C (5,131°F)

First identified: 1886

I n 1885, an unusual mineral (now known as argyrodite) was discovered in a silver mine near Freiberg in Germany. The mineralogist investigated it and found it contained 75 per cent silver and 18 per cent sulphur, leaving an unexplained 7 per cent. He realized this must be a new element that was in some respects metallic. Again, this was an element that Mendeleev had guessed must exist and had named eka-silicon in its absence. Mendeleev's predictions were especially accurate for this element: he was nearly spot-on in his guess of atomic weight (72 as opposed to the actual value of 72.6), was similarly close with the density, and correctly predicted that it would have a high melting point and be grey in colour.

A Miracle Cure?

Germanium compounds have inspired some fairly wild claims about their value when it comes to health – it is said, for instance, that the grotto at Lourdes (where a young girl reportedly had a vision of the Virgin Mary) has a high germanium content and that this might be a reason for the thousands of 'cures' that have been attributed to it. Germanium has also been promoted as a cure for AIDS, cancer and other ailments. However, there is no scientific evidence for these claims, and while it is present in trace elements in our diet (in garlic, for instance), too much germanium can damage your nervous system or kidneys.

There were initially no obvious uses for the metal, and it was only produced in fairly small quantities as it makes up less than two parts per million of the Earth's crust. However, during the Second World War, American researchers discovered it could be used as a semiconductor – one of several ways in which it is not obviously metallic, and one reason why we call it a 'metalloid' rather than a metal. (The metalloids are a much-debated group that tend to be identified by the fact that they can seem metallic or non-metallic in different allotropes. They are roughly grouped in the periodic table in a diagonal, from boron at the top left to polonium at the bottom right.*) It was later superseded by silicon and other substances, although it is now being used again as a semiconductor in solar panels. It is also used extensively in fibre-optic cables, as its high refractive index (a number that defines how light propagates in a medium) means it prevents the light escaping.

Arsenic

| 33 |
| As |
| 74.921595 |
| Arsenic |

Category: metalloid
Atomic number: 33
Colour: grey
Melting point: n/a

Sublimation point: 616°C (1,141°F)
First identified: ancient civilizations

ts compounds have been used to kill insects; as colouring agents; to preserve wood; for animal feed; as a medical treatment for syphilis, cancer and psoriasis; in fireworks; and (compounded with gallium) as a semiconductor. However, the name of arsenic will always be synonymous with its historic use as a poison.

---

* Boron, silicon, germanium, arsenic, antimony and tellurium are generally recognized as metalloids. Carbon, aluminium, selenium, polonium and astatine are also sometimes included.

For centuries before it became possible to identify its presence through analysis of hair samples in 1836, it was more or less impossible to detect arsenic poisoning from analysis of a victim's body, whether they had been given one large dose or slowly poisoned over time. It was known as 'inheritance powder' because of how often it was used to eliminate rich relatives. The Borgias are notorious for having used arsenic to amass their wealth – Pope Alexander VI and his two children Cesare and Lucrezia were responsible for the murder of numerous wealthy bishops and cardinals (whose property the pope would inherit after their death).

Alchemical Magic

The alchemists are often depicted as mad magicians who harboured weird beliefs about gold. In fact, they were merely the chemists of their time, struggling to understand what the world around them was made of – it can be easier to understand where their beliefs came from when you realize how many metals or metallic substances can be produced by truly astonishing processes. The writings of the thirteenth-century polymath Albertus Magnus include a description of how white arsenic (which looks like a white rock or sandy powder) could be mixed with olive oil and heated, producing the grey, metallic form of arsenic – as if by magic!

Arsenic was known to the ancient Egyptians, in the form of the yellow crystal sulphide compound called 'orpiment'. The Chinese used it as a pesticide at least 500 years ago, and Paracelsus (the alchemist also known as the father of modern toxicology) wrote of the preparation of metallic arsenic. Orpiment was used as a pigment in a historic type of paint known as Paris Green or Scheele's Green – it is known that Napoleon Bonaparte's apartment on St Helena during his final exile was decorated with wallpaper of this shade, which would give off arsenic gas when it became damp or mouldy, and it is speculated that this might have contributed to his death (although the evidence is not conclusive).

Mostly produced as a by-product of copper or lead refining, arsenic can come in various forms. Grey (or metallic) arsenic is a brittle semi-metallic solid, which is sometimes found in pure form, but it generally oxidizes to form arsenous oxide (giving off a pretty unpleasant garlicky smell in the process – that same smell could have been a tell-tale sign that someone had died of arsenic poisoning before better tests were available).

Selenium

Category: non-metal

Atomic number: 34

Colour: metallic grey

Melting point: 221°C (430°F)

Boiling point: 685°C (1,265°F)

First identified: 1817

T here are many elements that are both essential parts of the human diet and at the same time toxic if consumed in large doses. Selenium is one – it is crucial in the production of certain enzymes within the human body (we consume it in a wide variety of foodstuffs, including nuts and tuna). Recent clinical tests have suggested that the decline in selenium consumption (as a result of declining popularity of rich sources such as offal) is contributing to declining sperm counts in men; test subjects given a selenium supplement recorded considerably higher counts than a control group. However, if consumed in large doses, it can cause bad breath, hair loss, weakened nails, tiredness and confusion, and even fatal cirrhosis of the liver; though it is also (in the form of selenium sulphide) toxic to the scalp fungus that causes dandruff – and is used in safe doses in anti-dandruff shampoos.

Selenium was discovered by Jöns Jacob Berzelius in 1817. He investigated the chemical make-up of a reddish powder that was deposited in chambers in which sulphuric acid was being made. After initially mistaking it for tellurium (see page 112), he realized it must contain a new element, which he named after Selene, the goddess of the moon. (Berzelius discovered the hard way that selenium can cause bad breath, as he was affected by this himself after exposure to it!) Selenium can also take the form of a silvery, metallic substance, leading some chemists to define it as a metalloid.

Don't Try This at Home

This traditional warning before a science demonstration is one that Jöns Jacob Berzelius preferred to ignore. Many of his ground-breaking experiments were actually carried out in the kitchen of his apartment, which was on the corner of Nybrogatan and Riddargatan in the city of Stockholm.

The main contemporary use of selenium is as an additive to glass: it can either remove a greenish tint or be used to colour the glass reddish bronze, depending on how it is added. In various compounds, it can also be used in photovoltaic cells, solar cells and photocopiers, to create pipes (allied with brass) and to make synthetic rubber tougher.

Bromine

35
Br
79.904
Bromine

Category: halogen
Atomic number: 35
Colour: deep red
Melting point: −7°C
(19°F)

Boiling point: 59°C
(138°F)
First identified: 1826

romine is one of only a handful of elements that are liquids at normal atmospheric conditions. It is deep red, oily and toxic, with a nasty smell (the Greek word *bromos* means stench). It was discovered in 1826 by Antoine-Jérôme Balard; he took some brine, evaporated most of the liquid, then passed chlorine gas through it. When you do this, the bromine evaporates and can be collected, as an orange-red liquid, which Balard correctly guessed was an undiscovered element. This happens because salty water, and especially that found in the Dead Sea, contains bromides (a bromine atom that has acquired a negative charge).

Bromine was used far more widely a few decades ago – photography used the light-sensitive nature of silver bromide, potassium bromide was used as a tranquillizer, leaded petrol contained dibromomethane, and bromomethane (also known as methyl bromide) was used to fumigate soil. Some of these uses are now redundant, as better alternatives have been discovered, while others have been banned – the Montreal Protocol banning CFCs also called for a reduction in the use of many of these compounds, because bromine atoms damage the atmosphere too. It has, however, proved difficult to find replacements for some uses of bromomethane: in many places it is still used to kill pests in the soil, and to treat wood that is being transported. Bromine compounds continue also to be widely used as a fire retardant in plastic casings, such as those on laptop computers, and in flame-retardant substances in fire extinguishers.

Donning the Purple

Tyrian purple, a dye produced from the mucus excreted by the sea snail *Bolinus brandaris* (also known as the spiny dye-murex), was once a symbol of great wealth and power – the hardy, vivid dye was expensive to produce as it required thousands of snails to produce a tiny quantity. The magnificent purple togas worn by Roman emperors were dyed using it, and this is the origin of the phrase 'donning the purple', meaning 'taking power'.

Krypton

36	
Kr	
83.798	
Krypton	

Category: noble gas

Atomic number: 36

Colour: colourless

Melting point: –157°C (–251°F)

Boiling point: –153°C (–244°F)

First identified: 1898

B efore William Ramsay and Morris Travers discovered neon in 1898, they had in the same year already identified the fourth member of the group of noble gases, krypton. In this case, they liquefied and evaporated argon, to see if it would leave a heavier component behind. From 15 litres of argon they managed to produce 25 cm^3 of a gas: when they tested it with a spectrometer, it was clear that this was indeed a new element, which they named after the Greek word *kryptos*, meaning 'concealed', as it had been hidden in the argon.

Krypton is a gas without smell or colour and doesn't tend to react with any other element (apart from fluorine gas). It makes up only one part in a million of the Earth's atmosphere (by volume). It has been used to fill energy-saving fluorescent lightbulbs as well as in 'neon' lighting to expand the range of colours produced, and krypton fluoride has been used in the production of lasers.

Kryptonite Discovered!

In the 2006 movie *Superman Returns*, the chemical formula for kryptonite (Superman's nemesis) is given as 'sodium lithium boron silicate hydroxide with fluorine'. This is remarkably close to the mineral jadarite, which was discovered a year later – although those scientists who whipped up mass media attention by claiming that kryptonite had been found to be 'real' should have noted that the new mineral didn't contain fluorine, and it certainly didn't glow with an eerie green light. So, close, but no cigar.

In the Cold War, the radioactive isotope Kr-85 was used by Western scientists to spy on their counterparts behind the Iron Curtain – the isotope is produced at a fairly constant rate by nuclear reactors, and they realized that by estimating the amount of this isotope that was being produced by Western nuclear plants and subtracting this from the measurable total in the atmosphere, they could come up with a pretty good estimate of the level of nuclear activity in those countries that were in the Russian sphere of influence.

And, of course, krypton was the inspiration behind the planet Krypton, fictional home to Superman (not to mention Supergirl and 'Krypto the Dog') in the original DC Comics stories and the many films and comics that followed in its footsteps.

Rubidium

37
Rb
85.4678
Rubidium

Category: alkali metal

Atomic number: 37

Colour: silvery white

Melting point: 39°C (102°F)

Boiling point: 688°C (1,270°F)

First identified: 1861

Just as the noble gases (in the column on the right-hand side of the periodic table) share many properties, such as not being very reactive, the elements of group 1 on the left-hand side of the table are soft metals with low melting points, which are highly reactive. School chemistry teachers will sometimes demonstrate the strong reaction of lithium or sodium in water, but rubidium (which becomes liquid at a relatively low temperature) makes for a more dramatic and dangerous reaction – it can catch fire spontaneously in air and has to be stored in a vacuum or in a gas such as argon. When added to water, the explosion is instant, and so violent that the hydrogen being released from the water will often ignite.

This alkali metal has a handful of quite interesting applications: one of its isotopes is the radioactive rubidium-87. It has a half-life of 50 billion years: bearing in mind that it has only been about 14 billion years since the Big Bang, this is an extraordinarily slow rate of radioactive decay. It forms strontium-87 as it decays, which means that this property can be used to age ancient rocks accurately, by comparing the levels of rubidium and strontium using a spectrometer.

The Bunsen Burner and the Spectroscope

In the unlikely event that you are asked what these two brilliant scientific instruments have in common, the answer is Robert Bunsen: he invented the Bunsen burner, and was also (in 1859) the co-inventor (with Gustav Kirchhoff) of the spectroscope, which has been used to identify many new elements since. By 1861, the pair had used it to identify caesium and rubidium – the latter was found in the mineral lepidolite in 1861. It was named after the Latin *rubidus* (ruby) because of the vivid ruby-red lines that showed up on the mineral's spectrum.

Rubidium is also (along with caesium) used in atomic clocks, which align microwave radiation with the activity of electrons as they circle the atom and emit bursts of radiation. It is not naturally found in the human body, but it is harmless, as we can easily excrete it – it has been used to study the way that potassium moves around the body (since our bodies treat the two in the same way) and the radioactive isotope rubidium-82 has also been used to locate brain tumours.

Strontium

Category: alkaline earth metal	**Melting point:** 777°C (1,431°F)
Atomic number: 38	**Boiling point:** 1,377°C (2,511°F)
Colour: silvery grey	**First identified:** 1790

In the late eighteenth century, a strange rock was discovered in a lead mine near the village of Strontian, on the shores of Loch Sunart, in Scotland's western highlands. It was sent to Edinburgh for analysis, where the scientist Thomas Charles Hope proved it must contain a new element and commented on the fact that it causes a candle's flame to burn red. (Strontium, as it was named, was isolated by Humphry Davy in 1808.)

It is in the form of red flames that you are most likely to be familiar with the element – the red flares often seen at football stadiums are coloured using it, as are red fireworks. As a metal, it is similar to the other elements of group 2, including beryllium, magnesium and calcium: it is soft and reacts easily to form oxides. And it is only found in the form of mineral compounds – one of these is celestite (strontium sulphate), which was discovered in England's West Country in the eighteenth century (where it was being put to good use by local villagers as an ornamental gravel for their garden paths).

The radioactive isotope strontium-90 was produced by nuclear tests from 1945 onwards – it is a problematic isotope, as it can be absorbed into the food chain via grassland and dairy products, and our bodies mistake it for calcium, collecting it in bones and teeth. It is one of the harmful substances that was released in the 1986 Chernobyl accident and spread across parts of Russia and Europe.

Strontium's similarity to calcium has also led to its positive medical uses: in cancer therapy it is used as a radioactive tracer (allowing doctors to track cell movements and other processes in the body), and the non-radioactive salt, strontium ranelate, can be used to treat osteoporosis, as it slows the breakdown of old bone tissue and stimulates new bone production.

Yttrium

Category: transition metal

Atomic number: 39

Colour: silvery white

Melting point: 1,522°C (2,772°F)

Boiling point: 3,345°C (6,053°F)

First identified: 1828

Ytterby is a village on the Swedish island of Resarö, which is now mostly suburban housing (it is about half an hour by road from Stockholm). However, it used to be the home of the country's most productive mine, producing feldspar (for porcelain) and quartz. As it happens, it is also the place with the most elements named after it.

In 1787, Carl Axel Arrhenius (a soldier and amateur chemist) found a lump of black mineral that was fairly uninteresting to look at, but unusually heavy. It was later named 'gadolinite' (sometimes known as 'ytterbite'). The Swedish chemist Johan Gadolin (see page 129) established that 38 per cent of the rock must be a new, unidentified 'earth' (meaning that it was an unknown oxide), which proved impossible to reduce by burning with charcoal or other traditional methods.

It was only in 1828 that Friedrich Wöhler managed to rip the pure element out of the oxide, using a more violent reaction with potassium to separate out the oxygen, and producing pure yttrium. (The element has proved to be far more common on the moon than here – lunar astronauts have brought back significant amounts of it in moon rock.) However, as we will see, there were still three more unknown elements hiding in the gadolinite and, astonishingly, three more elements to be discovered as a consequence of these finds (see page 128).

Yttrium is a soft, silvery metal that has to be handled carefully, usually in nitrogen, as it combusts in air. It can be used to strengthen alloys of aluminium and magnesium, in microwave filters for radar technology, and in LED lighting and lasers. Yttrium oxide is added to glass to make it more resistant to heat and shock – for instance, it is used in bulletproof glass. There is also currently a lot of scientific interest in yttrium barium copper oxide (YBCO). In the 1980s, two American chemists showed that it becomes superconducting (meaning it will conduct electricity with no loss of energy) at the unusually high temperature of 95°

above absolute zero (–178°C). This could, in theory, enable us to build cheaper MRI scanners, as the YBCO could be kept as a superconductor, using liquid nitrogen rather than the more expensive liquid helium, but there are still technical difficulties to be solved in making this a reality.

Zirconium

Category: transition metal
Atomic number: 40
Colour: silvery white

Melting point: 1,855°C (3,371°F)
Boiling point: 4,409°C (7,968°F)
First identified: 1789

For over two millennia, we have known of a gold-coloured gemstone called *zargun* in Arabic, or zircon in English. Artificial versions of this gemstone are made today – they are more sparkly and dense than diamonds, but not as hard. Indeed, they were originally thought to be inferior diamonds – it was only in 1789, when Martin Klaproth managed to separate zirconia (or zirconium oxide) out from a zircon, that the element was discovered, and in 1824 that Berzelius managed to isolate some actual zirconium. The metal is hard, light, silvery and highly resistant to corrosion. (It also sparkles rather beautifully if you drop filings or dust into a Bunsen-burner flame, like a more spectacular version of the iron filings experiment your teacher may have shown you at school.)

91

The ceramics industry uses zirconium in pigments for glazing pottery, and, more importantly (in the form of zirconium oxide), in ultra-strong ceramics that will withstand high temperatures. A crucible made from it can be dipped into cold water while it is red-hot without damage. These ultra-strong ceramics are also used for knives, golf irons and cutting tools. The oxide is used, in addition, in the production of cosmetics, deodorants and microwave filters.

However, the most important use of zirconium is in nuclear reactors. As a metal that doesn't absorb neutrons, it is used in the cladding of nuclear fuel and other elements in nuclear reactors – however, zirconium can play an unfortunate part in nuclear accidents. At very high temperatures, the metal will react with steam, producing hydrogen (which explodes) and zirconium oxide (which causes the fuel rods that were previously contained in it to collapse) – this was part of what went wrong at Chernobyl in 1986. As the reactor heated up to unprecedented levels, the zirconium reaction created a loop in which the temperature became even more out of control, and the notorious accident was the result.

Niobium

Category: transition metal

Atomic number: 41

Colour: dull grey

Melting point: 2,477°C (4,491°F)

Boiling point: 4,744°C (8,571°F)

First identified: 1801

You may have noticed that there are particular periods of time in which the discovery of new elements accelerated. One was following the publication of Mendeleev's periodic table, as chemists started to search for the 'missing elements'. And earlier, at the end of the eighteenth century, Lavoisier's law of conservation of mass (1789) and Proust's law of definite proportions (1799) provoked John Dalton to formulate atomic theory and awakened a wider interest in the idea of 'elements'.

All this was part of the inspiration for the discovery of niobium – in 1801, the chemist Charles Hatchett started to ponder a specimen of the mineral columbite in the British Museum. His experiments led him to believe that it contained a new element, which he named columbium. Others subsequently doubted his findings, suggesting it was tantalum (which was separately discovered the following year). However, in 1844, the German chemist Heinrich Rose proved that columbite contained two elements (tantalum and niobium, which he named after

Niobe, the daughter of King Tantalus in Greek mythology). The pure metal was finally isolated in 1864.

It is a steely grey metal that is highly resistant to corrosion, as it forms a tough oxide layer. It is used in many alloys, especially stainless steel, and improves the strength of alloys that are to be used at low temperatures. It has also been used in rocket and jet engines, in oil rigs and gas pipelines.

American scientists continued to use the name columbium right up until 1950 – the controversy was only settled when an agreement was reached to take the European name for niobium, and the American name for tungsten (as opposed to the European 'wolfram'), though some US metallurgists, even today, still insist on using the name columbium.

It was tungsten (see page 137) that robbed niobium of a more prominent everyday role in our lives: niobium was the metal originally used for the filaments of incandescent lightbulbs, due to its high melting point, but it was soon replaced by tungsten, which melts at even higher temperatures.

Molybdenum

Category: transition metal
Atomic number: 42
Colour: silvery white

Melting point: 2,623°C (4,753°F)
Boiling point: 4,639°C (8,382°F)
First identified: 1781

For an element that most people know very little about, molybdenum is surprisingly crucial to human life. In plants and animals, there is a wide range of enzymes, one of which is nitrogenase. This is found in bacteria that live in the roots of plants such as beans. They absorb nitrogen from the air and emit ammonia: a key part of the process of 'nitrogen fixing', whereby nitrogen is converted into a form that humans and animals can digest. We rely on that nitrogen to manufacture proteins in our body. So, we simply couldn't survive without trace elements of molybdenum in the environment.

The element is also integral to the production of 'moly steel', an alloy of steel with molybdenum added to it. In the First World War, the earliest British tanks to be deployed on the Western Front were armour plated with three-inch manganese steel plates, which couldn't withstand direct hits. They were replaced with molybdenum steel plating, which was only one inch thick (so lighter) but proved to be far more effective. Moly steel is also used in high-grade construction, such as skyscrapers and bridges.

The name of the element comes from the Greek *molybdos*,

meaning 'lead': it doesn't occur naturally in its pure form and its main ore, molybdenite, was often mistaken for either a lead ore or graphite, which it closely resembles. Carl Wilhelm Scheele realized in 1778 that it was neither lead nor graphite. His friend Peter Jacob Hjelm successfully followed up on his work and isolated the element, a shiny, silvery metal, in 1781, using a method that looks like alchemical magic: he ground up some carbon with molybdic acid, then made a paste of this with linseed oil. When this was heated until it became red-hot, the metal emerged from the mixture.

Molybdenum is used today in tiny amounts in electric heater filaments, missiles and protective boiler coatings, and as a catalyst for refining petroleum. Molybdenum sulphides are also the main ingredient in lubricants that are heat-resistant at higher temperatures than petroleum-based oils such as WD-40.

Technetium

Category: transition metal
Atomic number: 43
Colour: silvery grey

Melting point: 2,157°C (3,915°F)
Boiling point: 4,265°C (7,709°F)
First identified: 1937

This is the substance that completes the story of Mendeleev's original four missing elements. After the discovery of scandium, gallium and germanium, there were many and varied attempts to identify and isolate element 43, which he had called eka-manganese. But these all failed until 1937, when the Italian scientists Carlo Perrier and Emilio Segrè at the University of Palermo in Sicily discovered it in an unexpected way. Segrè had visited the particle accelerator at Berkeley in America, and its creator Ernest Lawrence sent him a piece of molybdenum foil that had been bombarded with deuterons (nuclei from deuterium cells containing one proton and one neutron) in the cyclotron.

The scientists managed to isolate two radioactive isotopes of a new element, which they named technetium. This was a controversial discovery, as the creation of 'artificial' elements in this way was seen as a kind of cheating at the time, and the element was not widely recognized. We now know that all isotopes of the element are radioactive, which means that it is mainly formed in nuclear reactions in stars (and its relatively short half-life means that any technetium present at the formation of the Earth would have decayed and disappeared long ago).

However, the advances made in subatomic physics during the Second World War and the discovery of plutonium, which is similarly artificial, led to a revision in opinion: 'The Making of the Missing Chemical Elements', by Professor F.A. Paneth, was a key paper that persuaded the scientific community that there should be no distinction made between artificial elements and those that are naturally found, and also that the original discoverer of any isotope of an element should be allowed to name it. Perrier and Segrè soon issued a dignified response, suggesting for their discovery the name technetium, from the Greek for 'artificial', and it was finally accepted as a genuine element.

Ironically, natural reserves of technetium were discovered in 1972 – geologists found that there had been a spontaneous nuclear reaction long ago inside uranium deposits beneath Gabon in Africa, and there was a small amount of technetium still present.

The technetium produced today is gathered from spent nuclear fuel rods and is predominantly used (in the form of 'technetium-99m') for medical imaging: this isotope will bond with cancer cells and so can be used to identify the locations of tumours within the body.

Ruthenium

Category: transition metal

Atomic number: 44

Colour: silvery white

Melting point: 2,334°C (4,233°F)

Boiling point: 4,150°C (7,502°F)

First identified: 1844

Ruthenium is a shiny, silvery metal that is most often found with other platinum metals in minerals such as pentlandite and pyroxinite. (The six platinum metals are found in a rectangular group in the periodic table, and have similar properties: they are ruthenium, rhodium, palladium, osmium, iridium and platinum.) It is mostly mined along with nickel or platinum, but it is a rare metal on planet Earth – only about twelve tons of it are extracted every year.

Ruthenium was probably first identified by the Polish chemist Jędrzej Śniadecki in South American platinum ores: in 1808 he claimed to have discovered a new metal, which he wanted to call vestium. But when others tried to repeat his work, the metal wasn't found, so he gave up on the claim.

In 1825, the German chemist Gottfried Osann claimed to have found three new elements in platinum ores from the Urals: he named them pluranium, polinium and ruthenium. He was wrong about the first two, but Karl Karlovich Klaus at the University of Kazan confirmed the existence of ruthenium in the 1840s and kept Osann's name (which was based on the archaic name Ruthenia, an area that includes modern Russia).

Ruthenium is used in alloys, to make platinum and palladium harder, and is also alloyed with these and other metals in (for instance) jewellery, electrical contacts, solar cells and chip resistors. The renowned fountain pen, the Parker 51, uses a gold nib tipped with 96 per cent ruthenium (alloyed with iridium).

Rhodium

| 45 |
| Rh |
| 102.90550 |
| Rhodium |

Category: transition metal

Atomic number: 45

Colour: silvery white

Melting point: 1,964°C (3,567°F)

Boiling point: 3,695°C (6,683°F)

First identified: 1803

The streets of your local town or city would be much more unhealthy places if it weren't for rhodium, one of the rarest non-radioactive elements. Catalytic convertors in cars use a variety of oxidation and reduction processes to convert toxic gases and pollutants into cleaner emissions in exhaust gas. Palladium and platinum are also used for this purpose, but rhodium plays a key part in the conversion of 80 per cent of the nitrogen oxides into harmless nitrogen and oxygen.

It was discovered by William Wollaston in 1803 – he was collaborating with Smithson Tennant on an attempt to purify platinum for sale. After they dissolved platinum in the fierce acid blend known as *aqua regia*, Smithson Tennant investigated the residue left behind (see pages 141–3) while Wollaston continued working with the solution of platinum. He removed the platinum and then palladium (see next entry) by precipitation, which is the creation of a solid from a solution. He was left with sodium rhodium chloride, in the form of beautiful rose-red crystals, which inspired the name rhodium, from the Greek *rhodon*, for 'rose-red'. He went on to extract the metal itself from this compound.

Rhodium is used in alloys with platinum for thermocouples, as a coating for optic fibres, and as an electrical contact material. As well as its role in catalytic convertors, the chemicals industry uses it as a catalyst in a variety of other ways. You might, for example, encounter it on a regular basis in mint flavouring. Such flavouring was originally derived from mint plants, but the Nobel Prize-winning Japanese chemist Ryoji Noyori devised a high-quality method of menthol production that relies on rhodium catalysis.

Green Chemistry

Ryoji Noyori is an inspirational example of how chemists can impact on the world positively. He is an advocate of green chemistry: the design of sustainable products and processes that minimize the use of hazardous substances. In one recent article, he argued that 'our ability to devise straightforward and practical chemical syntheses is indispensable to the survival of our species'.

Palladium

Category: transition metal

Atomic number: 46

Colour: silvery white

Melting point: 1,555°C (2,831°F)

Boiling point: 2,963°C (5,365°F)

First identified: 1803

razilian miners in the eighteenth century occasionally found lumps of *ouro podre*, meaning 'worthless gold'. This was a naturally occurring alloy of gold and palladium – the latter would also occasionally be found on its own, but it was not identified as an element until 1803. At the same time that William Wollaston discovered rhodium, precipitating it from platinum, he also produced a quantity of palladium. His immediate response to this was not to announce it in the usual manner to the scientific community.

Instead, he wrote a pamphlet eulogizing his discovery and put the metal up for sale at a shop in Gerrard Street, Soho. The leaflet was called 'Palladium; or, New Silver', and it offered this 'In Samples of Five Shillings, Half a Guinea & One Guinea each'. It was only when a chemist challenged his claims, suggesting it was merely a platinum alloy, that he published his method.

Like rhodium, palladium is used in catalytic converters in cars; it is good at reducing unburned or partially burned hydrocarbons before they are emitted into the atmosphere. It is also widely used in the electronics industry; for instance, in ceramic capacitors (which are made of thin layers of ceramic and palladium). *Ouro podre* is prized more highly today than it was in the past – 'white gold' is a popular option for jewellery and is most often made from a gold and palladium alloy.

Palladium has more than once been hailed as providing a potential solution to our energy problems. In 1989, the American scientists Martin Fleischmann and Stanley Pons claimed they had electrolysed heavy water using a platinum anode and a palladium cathode and created energy from a nuclear fusion reaction. This would have been an amazing breakthrough, but it proved illusory, as the result could not be replicated.

Second, palladium could theoretically solve the problem of large-scale hydrogen storage (for future fuel cells). It has a weird quality: hydrogen molecules are absorbed into its structure, where they diffuse into the metal and are compressed into a space up to a thousand times smaller. Palladium is currently too expensive for this to be a useful property, but if a way could be found to create a cheaper compound, alloy or palladium product that behaves similarly (as a 'hydrogen sponge') it could be extremely useful in future.

Silver

Category: transition metal

Atomic number: 47

Colour: silver, of course

Melting point: 962°C (1,764°F)

Boiling point: 2,162°C (3,924°F)

First identified: prehistory

S ilver is a native metal that can naturally be found in pure form, so we have known about this beautiful shiny metal for 10,000 years or more. It was probably first extracted from ores in the area of ancient Turkey and Greece by the Chaldeans in about 3000 BC using 'cupellation'. This is a method whereby molten metal ores (containing, for instance, lead-zinc, copper or copper-nickel) are heated in a cup over which air is blown – the other more reactive metals are thus oxidized and the molten silver is isolated.

It was used from around this time for coins, as well as for luxury household items. It has often been strengthened by the addition of copper or other metals, in which case it is known as sterling silver. Unlike its precious-metal cousin, gold, silver does tarnish – it gradually becomes darker and duller as it reacts with sulphur to form silver sulphide on the surface, which has to be polished off regularly. But it has nonetheless been valued highly throughout history.

If you're fond of either mirrors or selfies, silver has played a key role in the development of these ways of admiring yourself. Due to its reflectivity, it was historically the main metal used in making the backs for mirrors (although the cheaper aluminium is usually used now). In 1727, Johann Heinrich Schulze (a German scientist) made a slurry of chalk and a silver nitrate salt and found that it was blackened by the light. After experimenting by making images with stencils, he had created the origins of the science of photography – the separation and combination of ions on the surface of the mixture was forming darker and lighter areas (although Schulze was unable to fix and keep this image). In 1840, Henry Fox Talbot worked out how to fix the image on paper coated in silver iodide using gallic acid, and all the necessary pieces were in place.

Of course, modern selfies are digitally created without the need for the original chemical processes, but it's worth taking a moment to appreciate what a magical discovery it was at the time. And even in our modern age, silver is finding new applications – it is now possible to scroll through the images on your phone wearing gloves that have silver thread sewn into the fingertips, so that the touchscreen can be used without you getting cold hands.

Cadmium

48
Cd
112.414
Cadmium

Category: transition metal

Atomic number: 48

Colour: silvery blue

Melting point: 321°C (610°F)

Boiling point: 767°C (1,413°F)

First identified: 1817

admium is a toxic substance, which can cause birth defects and cancer, as well as the disease that the Japanese call *itai-itai* (meaning 'ouch-ouch' because of the joint pain it causes). There was a notorious outbreak of *itai-itai* in the 1960s in the Jinzu River basin, when rice crops in the area were affected by pollution from a local zinc mine. Our bodies do have some natural defences against cadmium poisoning, but above a certain threshold it can be very dangerous. Cadmium is often produced as a by-product of zinc, to which it has many similarities. It is also similar in some respects to mercury, and all three elements are in a vertical line in the periodic table.

It was identified in 1817 after German apothecaries (or pharmacists) who were making zinc oxide by heating a natural zinc carbonate called cadmia found that the oxide was sometimes discoloured rather than pure white. Friedrich Stromeyer, who was Inspector of Pharmacies at the time, investigated this and separated out a brown oxide, which he then reduced to a new metal by heating it with carbon.

Cadmium has had many uses over the years: for instance,

it was used to make the pigment cadmium yellow (one of Monet's favourite paints) and it could also be produced in brown, red and orange with the addition of other substances such as sulphur or selenium. It was used further (in the form of cadmium sulphide) to give casserole dishes their lovely orangey-red colour.

Old batteries and cathode ray tubes in early colour TV sets contained cadmium. And it is still used today in rechargeable nickel-cadmium batteries. It also absorbs neutrons, so has found uses in nuclear reactors. And many components in heavy machinery, such as oil platforms, contain cadmium. Typically, we are trying to replace most uses of this toxic element, which makes it all the more extraordinary that people still voluntarily inhale it in cigarette smoke, and litter the environment with it when they fail to dispose safely of nickel-cadmium batteries.

Indium

Category: post-transition metal
Atomic number: 49
Colour: silvery grey

Melting point: 157°C (314°F)
Boiling point: 2,072°C (3,762°F)
First identified: 1863

Indium tin oxide is an extremely useful compound: it is transparent to visible light, electrically conductive, and it bonds strongly to glass. This combination means it can be used in flat-panel displays (LCD screens on computers and other devices – the acronym stands for Liquid Crystal Display). It allows individual pixels to receive signals without light from other pixels being affected.

This has led to a sharp rise in the price of indium in recent years; it is a relatively rare metal, comprising only about 0.1 parts in a million of the Earth's crust (so it's about as rare as silver). It previously had few uses, the global production in 1924 being a couple of grams, but now we use over 1,000 tons a year (with half of that coming from recycled sources). Suggestions have been made that the supply of indium might run out over the next decade. This may, though, be slightly alarmist – there is still a lot of the metal available, and rising prices generally mean that suppliers become more ingenious in their extraction techniques.

Indium is a moderately toxic, soft, silvery metal that is unusually sticky (it is used as a solder for this reason; in its pure form, it sticks tightly to other metals) and has many high-tech applications. When you bend a piece of the metal, it gives out a 'cry' similar to the noise tin makes – it's actually more of a crackle and is the sound of the molecules rearranging themselves. It is workable at low temperatures, so it has been used in cryogenic pumps and equipment designed to be used at temperatures close to absolute zero. In alloys it can make significant differences to other metals – for instance, a gold and indium alloy is much harder than gold. The compounds, indium gallium arsenide and copper indium gallium selenide are also used in solar cells.

The metal is named after the colour indigo – when the German chemist Ferdinand Reich discovered it in minerals that were rich in zinc, the tell-tale sign of a new element came when his colleague Hieronymus Richter saw a bright indigo line on the atomic spectroscope. (The two men later fell out because Richter had been claiming that he alone had discovered it.)

Tin

Category: post-transition metal

Atomic number: 50

Colour: silvery white

Melting point: 232°C (449°F)

Boiling point: 2,602°C (4,716°F)

First identified: ancient civilizations

Tin was a hugely important element in our history. It is a soft metal with a low melting point, which doesn't corrode through oxidation. It was already in use about 10,000 years ago, but a key breakthrough came when metal workers worked out how to combine tin and copper to produce bronze in about 3500 BC. Metals had been smelted and extracted before this date, but this was the first significant alloy. It became clear that bronze combined the advantages of its two constituent metals – it was harder than tin, but melted at a lower temperature than copper, which made it easier to work with. Tin would also be alloyed with lead, copper and antimony as pewter, and used to 'tin-plate' ironware to stop it rusting.

The metal was a hugely important economic resource for

the next few millennia, mined at various sites around the Mediterranean and in Cornwall, which may have been one of the motivations behind the Roman invasion of Britain. (The abbreviation 'Sn' comes from the Latin *stannum*.)

While it doesn't oxidize, pure tin suffers from 'tin pest', a process by which it decays gradually into a grey, powdery dust. This starts happening at temperatures below 10°C but becomes much more severe at about 30°C below zero. It has been blamed for at least two historical disasters: during Napoleon's failed Russian campaign in 1812 it is said that the buttons on his soldiers' uniforms started to turn to dust in the cold winter, exacerbating their problems with hypothermia. And when Captain Scott and his companions set off back from the South Pole (having been beaten there by Amundsen), they returned to a store of tinned provisions but found that some paraffin had leaked through tiny holes in the tin – they all died of exposure subsequently.

'Tin cans' and 'tin foil' are actually made of aluminium (although they can be lined with tin), but many bells and organ pipes are still made of tin (usually alloyed with lead) and it is a significant ingredient in many alloys; for instance, those used for soldering (joining pieces of metal together). It is also used in glass production – the molten glass is floated on a bed of molten tin to create a flat surface. And many children continue to enjoy playing with tin toys (which are now mostly manufactured in China).

Antimony

Category: metalloid
Atomic number: 51
Colour: silvery
Melting point: 631°C (1,167°F)

Boiling point: 1,587°C (2,889°F)
First identified: approx. 1600 BC

A ntimony has been used in various forms for at least 5,000 years. An ancient fragment of an antimony artefact from the Sumerian civilization was found in the nineteenth century in what is now Iraq – it has sometimes been described as having been part of a vase, although this seems unlikely, as antimony is too brittle to be shaped into a vessel. The mineral stibnite (a black form of the pigment antimony sulphide) was used as mascara (known as khol) by the Egyptians by about 1600 BC. (The abbreviation Sb comes from *stibium*, the Latin name for stibnite.) In the biblical description of that infamous bad girl, Jezebel, she is said to have 'painted her eyes with kohl and dressed her hair' in a final act of defiance. The pigment yellow lead antimonite was also used by the Babylonians to glaze ornamental bricks.

In medieval times, antimony was best known for its supposed medicinal qualities: it is a rather toxic substance, but it was used as an emetic and as a laxative. The 'Antimony War' of the seventeenth century was a furious debate following the esteemed physician and alchemist Paracelsus's enthusiasm for it. One German writer, masquerading as a monk called 'Basil Valentine', advocated its many uses in a book called *The Triumphal Chariot of Antimony*. He admitted it was poisonous, claiming the name came from 'anti-monk' – as it was known for poisoning monks – but at the same time claimed he could make a non-toxic version using alchemy. It has also been suggested that medicines made from the element may have killed the composer Mozart.

Antimony is mainly used today in the electronics industry (for instance, in semiconductors, diodes and, alloyed with indium, infrared detectors). It can be alloyed with soft metals such as lead to make them harder; was traditionally used in pewter, where the copper and antimony together hardened the tin and lead; and an alloy of lead and antimony can be used to make bullets or type metal (as found in old-fashioned printing presses). It is also used in a variety of flame-retardant materials such as paints and enamels.

Tellurium

Category: metalloid
Atomic number: 52
Colour: silvery white
Melting point: 449°C (841°F)

Boiling point: 988°C (1,810°F)
First identified: 1783

Tellurium is an element for which the future supply is uncertain. Traditionally, it was used in alloys – for instance with copper, lead and stainless steel – to make them harder, more workable or resistant. It can also be used to vulcanize rubber and to tint glass. But the real area of growth in demand has come from its use in rewritable CDs and DVDs and in the manufacture of solar panels, because it can capture energy efficiently in the form of the compound cadmium telluride.

The problem is that tellurium is obtained as a by-product of copper production – it is recovered from the 'anode slime' produced in the electrolytic refining process. Copper production has declined in recent years, and there have also been changes in the processes used (as different types of copper are extracted), and this has affected the supply of tellurium, pushing prices up considerably.

Tellurium is usually produced as a dark-grey powder, though as a metalloid it also has a shiny, silver metallic form. It is somewhat toxic – like selenium – and, when handled, it can give you a nasty case of garlic breath, as well as blacken your hands rather unpleasantly. Appropriately enough, it was discovered in Transylvania: the Austrian mineralogist Franz Joseph Müller von Reichenstein found a shiny ore, which turned out to be gold telluride rather than antimony or bismuth, as he had suspected. He proved it contained a new element, but his work wasn't widely recognized until he sent a sample to German chemist Martin Klaproth, who agreed his findings were correct.

The planet Uranus had recently been discovered. Mindful of an ancient tradition that connected each of the seven known celestial bodies with a metal (for instance, the sun was associated with gold, the moon was supposed to nourish silver ores in the ground, Mars was connected with iron, and so on), Klaproth united the Ancient Greek *tellus* (for 'Earth') with the first part of Uranus's name to come up with the term 'tellurium'.

Iodine

53
I
126.90447
Iodine

Category: halogen
Atomic number: 53
Colour: black (with a purple gas)

Melting point: 114°C (237°F)
Boiling point: 184°C (364°F)
First identified: 1811

I n the past, people were prone to suffering from goitres – a pronounced swelling in the neck – in specific areas of land. The same areas tended to experience a high rate of learning disabilities (in particular, the condition that unfortunately used to be known as 'cretinism'). These areas were often inland and far from the sea, so early medical writers suspected there might be a connection.

The medieval physicians Galen and Roger of Salerno suggested treating goitres with marine sponges and seaweed respectively – there is similar advice in Chinese texts of the period. And Paracelsus suggested that there might be a mineral in seawater that played a role in preventing the condition. They were all on to something, but the connection only became clear more recently. In 1811, the French chemist Bernard Courtois was making saltpetre (potassium nitrate) using seaweed ash for the potassium. After adding sulphuric acid, he was surprised by the appearance of purple fumes, which condensed to form black crystals. It soon became clear that he had witnessed a new element, iodine (whose name derives from the Greek for violet, *iodes*) as both a gas and then a solid. (Incidentally, it is widely believed, even by chemists, that iodine sublimates and doesn't have a liquid phase – in fact, it has a fairly narrow temperature range in which it is a stable liquid, but it can rapidly pass from solid to gas when heated.)

Iodine is quite toxic and explosive, so needs to be treated carefully. However, it has commercial uses – in the early photographic form of daguerreotypes, and more recently in disinfectants, animal feeds and inks and dyes. Like other halogens, it can form a stable ion, called iodide, which is widely found in seawater in compounds, such as potassium iodide. This is how it passes safely into the human food chain, via seafood and plants that absorb it from the ocean spray.

And this is how the ancient mystery was solved. After some rather dangerous early experiments with iodine, it was discovered that potassium iodide was effective at treating goitres. And the condition of cretinism was mostly eradicated in the developed world once safe quantities of iodine started to be added routinely to commercially available salt. Iodine is needed by the thyroid gland – too much or too little can cause medical conditions, such as iodine deficiency, which remains a problem in many developing countries.

Xenon

Category: noble gas

Atomic number: 54

Colour: colourless

Melting point: –112°C (–169°F)

Boiling point: –108°C (–163°F)

First identified: 1898

F or Sir William Ramsay and his colleague Morris Travers, 1898 was a remarkable year. It saw them, having discovered argon four years earlier, identifying both krypton and neon through their continued experiments with air. But they weren't done yet – the industrial chemist Ludwig Mond gave them a liquid air machine, and they continued experimenting. On 12 July, they were using a vacuum vessel to remove some residues of argon and krypton when they noticed a small bubble of gas remaining behind. They treated this with potassium hydroxide to remove any carbon dioxide and ended up with a tiny sample in a vacuum tube – this gave off a beautiful blue glow when heated, and rendered a spectroscope reading quite different to krypton. The two men concluded they had found a new element and (once it became apparent that all the words meaning 'blue' had already been used) named it xenon, after the Greek word for 'stranger'.

Xenon is a heavy gas. It is fascinating, for instance, to see how rapidly a balloon filled with it falls to the floor. (This needs to be done with care though, as it is an expensive gas to produce.) For a long time it was believed that xenon was completely inactive, until the British chemist Neil Bartlett and his team, working in Canada, came up with a brilliant experiment that showed it could form a compound with fluorine and platinum. Subsequently, more compounds have been found – it will react in the right conditions with gold, hydrogen and sulphur. However, the compounds are all unstable, as they tend to oxidize easily.

As a result, the main uses of xenon are currently based on the pure element. It is used in car headlamps to provide instant illumination, in high-speed electronic flash bulbs for photography, and in some kinds of lasers and sunbeds. It can theoretically be used as an effective anaesthetic, like a kind of laughing gas, but currently this would prove too expensive to be viable. Finally, Xenon Ion Propulsion Systems (XIPS) may sound like something from a science-fiction novel, but the term refers to a real piece of technology that is employed to manoeuvre satellites. The system ionizes xenon atoms and accelerates them to a speed of about 20 miles per second before ejecting them, propelling the satellite through space.

Caesium

Category: alkali metal
Atomic number: 55
Colour: silvery gold
Melting point: 28°C (83°F)

Boiling point: 671°C (1,240°F)
First identified: 1860

C aesium would be a lovely element to play with if it weren't so extraordinarily reactive – it becomes liquid just above room temperature, so can be melted with the warmth of your hands. It is one of only three gold-coloured metals (along with copper and, of course, gold), although the gold colour disappears in 100 per cent pure samples, as it is caused by tiny traces of oxygen. The problem is that it is extremely reactive in air, so it has to be stored under oil or in an inert gas like argon, and if you drop it in water, it is even more explosive than its fellow alkali metals: lithium, potassium, sodium and rubidium. (For an explanation of why caesium is also slightly more reactive than the heavier alkali metal, francium, see page 160.)

You may remember that Robert Bunsen and Gustav Kirchhoff used the spectroscope invented by Kirchhoff to discover rubidium in 1861. At that point they had already discovered caesium – they had been studying a sample of mineral water when they noticed unexpected blue lines in the spectrum, indicating a new element, which they named after the Greek word for 'sky blue'. They were able to produce caesium chloride, but it would be another twenty-two years before a sample of caesium was isolated from molten caesium cyanide (by Carl Theodor Setterberg at the University of Bonn).

Just a Second

If you're ever asked to give the internationally recognized definition of a second, just memorize this: in 1967, the official definition was set by the International Bureau of Weights and Measures at 9,192,631,770 cycles of the radiation that gets an atom of caesium-133 to vibrate between two energy states. Simple!

Caesium compounds are used as drilling fluid and in the production of optical glass. But the most significant use of this element is in the caesium clock – it has replaced rubidium as most commonly used for this. The principle behind atomic clocks is the frequency of a change of energy levels caused by the magnetic field of the nucleus, and the stable isotope caesium-133 is currently the best candidate (although, as well as rubidium, strontium could also be used and, in theory, ytterbium).

Barium

Category: alkaline earth metal

Atomic number: 56

Colour: silvery gold

Melting point: 729°C (1,344°F)

Boiling point: 1,845°C (3,353°F)

First identified: 1808

I f you've ever been unlucky enough to have a barium meal or enema, you probably won't have very happy memories of it. As a heavy element, it shows up clearly on X-rays, so it is often used to diagnose diseases that affect the intestine or the oesophagus. A suspension of barium sulphate (barite) in fluid is administered, with flavouring such as strawberry or peppermint added in a vain attempt to make it palatable. The advantage of barium sulphide is that it is insoluble in water, so it will completely leave the digestive system afterwards. Soluble barium salts are quite poisonous, so this is crucial. Barium carbonate has been used as rat poison, while barium acetate, stolen from a chemistry classroom, was employed as a murder weapon by Texas teenager, Marie Robards, who used it in 1993 to kill her father.

Barium sulphate is naturally found as a white, slightly transparent ore, an unusually heavy rock. In the seventeenth century, a shoemaker from Bologna called Vincenzo Casciarolo discovered a rock that would glow in the dark if it were heated sufficiently during the day. He was excited by the idea that this stone (called *lapis solaris* or Bologna Stone) might somehow be a way to manufacture gold, but it turned out merely to be barite.

Barium is only found in compounds, as it is highly reactive in air – it can, for instance, be found in the ore, witherite (barium carbonate). Attempts to extract the element from these ores by smelting it with carbon failed, but Humphry Davy succeeded in isolating the soft, greyish metal in 1808 by electrolysing barium hydroxide. It is named after the Greek *barys*, meaning 'heavy'. Also known as 'heavy spar', barite is used in the petroleum industry as a weighting agent in the drilling mud used in creating oil wells.

It has a few other uses, such as in making paint and in glassmaking, and barium nitrate is used for the green colour in fireworks. The fascinating compound YBCO (yttrium barium copper oxide; see page 89), which can be used as a superconductor at relatively high temperatures, is attracting a lot of scientific interest. And barite has an interesting role for scientists studying the ocean: because it is insoluble and remains stable for millions of years, the accumulation of this ore in ocean sediments can give us crucial information about how productive marine phytoplankton were in particular periods of the planet's past.

The Lanthanides

The lanthanides are a sequence of elements that are gathered together in a single strip below the rest of the periodic table. They are also (together with scandium and yttrium) known as the 'rare earths' because these chemicals were all isolated as oxides (also known as earths) and they derive from rare minerals (although the elements themselves are mostly not that rare). They are grouped together in a single spot in the main table because of their similarity and also because of a quirk in the way their electrons are arranged: they all have the same number of electrons in their outer orbit, which is the orbit that reacts with other atoms, and thus defines the chemical properties of the element. Each separate lanthanide has a different number of electrons, but as the atomic number goes up, the extra electron is added to an inner orbit and the same set of three electrons remains in the outer shell, so they all have similar properties.

It is unwise to suggest to a chemist that any particular element is 'boring' – you'll probably find they spent years working on a PhD thesis concerning some minor property of that exact element – but it would be somewhat repetitive to give every lanthanide equal treatment, so, instead, here is a table of their key information, with a summary of key points about each below.

Lanthanum (La) Atomic number 57; Melting point: 920°C (1,688°F); Colour: silvery white; Boiling point: 3,464°C (6,267°F); First identified: 1839

Cerium (Ce) Atomic number 58; Melting point: 795°C (1,463°F); Colour: iron grey; Boiling point: 3,443°C (6,229°F); First identified: 1803

Praseodymium (Pr) Atomic number 59; Melting point: 935°C (1,715°F); Colour: silvery white; Boiling point: 3,529°C (6,368°F); First identified: 1885

Neodymium (Nd) Atomic number 60; Melting point: 1,024°C (1,875°F); Colour: silvery white; Boiling point: 3,074°C (5,565°F); First identified: 1885

Promethium (Pm) Atomic number 61; Melting point: 1,042°C (1,908°F); Colour: silver; Boiling point: 3,000°C (5,432°F); First identified: 1945

Samarium (Sm) Atomic number 62; Melting point: 1,072°C (1,962°F); Colour: silvery white; Boiling point: 1,794°C (3,261°F); First identified: 1879

Europium (Eu) Atomic number 63; Melting point: 826°C (1,519°F); Colour: silvery white; Boiling point: 1,529°C (2,784°F); First identified: 1901

Gadolinium (Gd) Atomic number 64; Melting point: 1,312°C (2,394°F); Colour: silver; Boiling point: 3,273°C (5,923°F); First identified: 1880

Terbium (Tb) Atomic number 65; Melting point: 1,356°C (2,473°F); Colour: silvery white; Boiling point: 3,230°C (5,846°F); First identified: 1843

Dysprosium (Dy) Atomic number 66; Melting point: 1,407°C (2,565°F); Colour: silvery white; Boiling point: 2,562°C (4,653°F); First identified: 1886

Holmium (Ho) Atomic number 67; Melting point: 1,461°C (2,662°F); Colour: silvery white; Boiling point: 2,720°C (4,928°F); First identified: 1878

Erbium (Er) Atomic number 68; Melting point: 1,529°C (2,784°F); Colour: silver; Boiling point: 2,868°C (5,194°F); First identified: 1842

Thulium (Tm) Atomic number 69; Melting point: 1,545°C (2,813°F); Colour: silvery grey; Boiling point: 1,950°C (3,542°F); First identified: 1879

Ytterbium (Yb) Atomic number 70; Melting point: 824°C (1,515°F); Colour: silver; Boiling point: 1,196°C (2,185°F); First identified: 1878

Lutetium (Lu) Atomic number 71; Melting point: 1,652°C (3,006°F); Colour: silver; Boiling point: 3,402°C (6.156°F); First identified: 1907

The Lanthanides' General Properties

Most of the lanthanides are silvery metals, soft enough to be cut with a knife. Lanthanum, cerium, praseodymium, neodymium and europium are all highly reactive, rapidly forming an oxide coating. The other lanthanides are all prone to corrosion if mixed with other metals, and to becoming brittle if they are contaminated by nitrogen or oxygen. They react more rapidly with hot water than cold and produce hydrogen in the reaction. They tend to burn fairly easily in air. The lanthanides are almost always found in the two minerals monazite and bastnäsite – they tend to be mixed together in fairly steady proportions in these (with about 25–38 per cent being lanthanum) and increasingly small amounts of the lanthanides with higher atomic numbers, which are heavier and so sank deeper into the Earth's mantle in the volatile past.

Lanthanum

Lanthanum was discovered in 1839 by Swedish chemist Carl Gustav Mosander, and isolated in 1923. As an alloy it shares palladium's ability to act as a 'hydrogen sponge', absorbing the gas at high density – although lanthanum is probably too heavy for this property to have any commercial value. As 25 per cent of the alloy 'mischmetal' (which also contains 50 per cent cerium, 18 per cent neodymium and assorted other lanthanides), it is used in flints in cigarette lighters. It can neutralize phosphorus, so is used in ponds to prevent the unwanted growth of algae.

Cerium

Cerium was discovered by Jöns Jacob Berzelius and his colleague Wilhelm Hisinger in 1803. Whereas most of the lanthanides were found together in monazite or bastnäsite, cerium was found separately in cerium silicate, a cerium salt. The most common lanthanide in the Earth's crust, it has some nice environmentally friendly uses: it produces a red pigment, for example, that is much safer than those obtained from cadmium, mercury or lead in paints; and a small amount added to fuel can reduce the number of polluting particulates produced in exhaust. It is also used to coat the walls of self-cleaning ovens, converting cooking residues to an ashy substance that is relatively easily wiped away. If you file or scrape filings of cerium off a block, they spontaneously combust – a property known as 'pyrophorism'.

Praseodymium

When Carl Gustav Mosander discovered lanthanum, there was a residue left behind that he suspected was another element, which he called didymium. In 1885, the Austrian chemist Carl Auer von Welsbach finally showed that this was a mixture of (mostly) two elements – praseodymium and neodymium. The main use of praseodymium is in aircraft engine parts as part of a high-strength alloy with magnesium. Like other lanthanides, it is used in carbon arc electrodes for studio lights. It can also be used to give glass and enamel a strong yellow colour, and to produce the glass in welders' goggles, where it filters out yellow light and infrared radiation.

Neodymium

Neodymium – the other part of Mosander's 'didymium' – was isolated in 1925. Its most important role is in extremely strong 'NIB' magnets (made from an alloy of neodymium, iron and boron). These are widely used in the magnets that power motors in electric cars. Neodymium is also used to make the glass in welder's goggles, and in sunbeds, where it transmits tanning UV, but not infrared, rays.

Promethium

Most elements with an atomic number lower than that of bismuth (the chemical element of atomic number 83: a brittle, reddish-tinged grey metal) have a stable form. The two exceptions are technetium (see page 96) and promethium, whose isotopes have a half-life of eighteen years at most. As a result, promethium no longer occurs naturally on Earth (although large amounts are being manufactured by a star in the Andromeda system, for unknown reasons). Its first confirmed discovery was in the fission products of uranium fuel taken from a nuclear reactor in 1945. It can also be created by bombarding neodymium and praseodymium in a particle accelerator. Briefly used to replace radium in the luminous dials of watches, it is only really utilized today for research purposes. It affords, however, another example of the periodic table being used to find a 'missing element', as it was predicted by the Czech chemist John Bohuslav Branner in 1902, and also by Henry Moseley (see page 3) after he rearranged the periodic table in 1913, that the gap between neodymium and samarium would eventually be filled.

Samarium

Samarium was the first element named (indirectly) after a person. Colonel Samarsky, a Russian mine official, granted mineralogist Gustav Rose access to some samples, one of which turned out to be a new mineral, which, in gratitude to the colonel, Rose called samarskite. In 1879, Paul-Émile Lecoq de Boisbaudran (discoverer of gallium) extracted didymium from the mineral, but he also managed to extract a new element, which he called samarium. It has some specialized uses in lasers, glass production and lighting. When alloyed with cobalt it makes strong magnets, although these have been superseded by NIB versions.

Europium

Europium was isolated and named by French chemist Eugène-Anatole Demarçay in 1901, having been identified by several different scientists independently. Its most useful properties are connected to phosphorescence: phosphors are the substances used to produce a glow when stimulated by electrons in, for instance, traditional televisions. Red phosphors used to be the weakest – but when they are 'doped' with a small amount of europium, the output light is much stronger. Europium is also used (in a gas combination) in white fluorescent lightbulbs – and it is the key element involved in making a phosphorescent anti-forgery mark on euro banknotes.

Gadolinium

Like samarium, gadolinium was extracted from a sample of didymium by Paul-Émile Lecoq de Boisbaudran – this was in 1886, six years after an oxide of the element had been identified by Swiss chemist Jean Charles Galissard de Marignac. It is often used in alloys; for instance, to make iron and chromium easier to work with. It is the best-known neutron absorber among the elements, so is used in nuclear reactors. It is also used in MRI scanning, where an injected gadolinium compound can enhance the image produced.

Terbium

We've seen how yttrium came from a mine near the Swedish village of Ytterby. Three more elements would be discovered directly in the ores taken from the area and named after the village, in one of the most confusing and boring pieces of naming in the history of the periodic table: these were erbium (1842), terbium (1842), and ytterbium (1878). And to make things worse, three more elements owed their name indirectly to the village: holmium (1878) was named after the Swedish capital Stockholm, thulium (1879) after Thule, the mythical name for Scandinavia, and, just for a bit of variation, gadolinium was named after Johan Gadolin, who had identified the mineral from Ytterby containing yttrium in the first place.

Confused yet?

Anyhow, terbium is mainly used in compounds in solid-state devices, in low-energy lightbulbs, X-ray machines and lasers. Its most interesting use is in the alloy of terbium, dysprosium and iron, which flattens out in a magnetic field. This property can be used to create loudspeakers that you attach to a flat surface, such as a window pane, turning that flat surface into an amplifier of the sound.

Dysprosium

Dysprosium was discovered as what at first appeared to be an impurity in another lanthanide – in this case erbium. Over a period of years, it was shown that the impurity contained, as well as dysprosium, at least two more separate elements: holmium and thulium. It took some exceptionally patient and repetitive experiments by Paul-Émile Lecoq de Boisbaudran to isolate dysprosium, so he named it after the Greek word *dysprositos*, meaning 'hard to get'. It can be used in nuclear reactor control rods, as it absorbs neutrons well. More importantly, it is added to alloys to make neodymium-based magnets designed to be used at high temperatures, at which it retains its magnetism well. Such magnets can be used in electric cars and wind turbines, both of which represent growing markets. Dysprosium, then, faces potential supply problems in the years ahead – it is the most expensive lanthanide and remains as hard to acquire today as it was when de Boisbaudran first named it.

Holmium

Holmium's main practical role is in high-performance lasers used to vaporize certain types of tumours with minimal damage to the surrounding tissue – these need yttrium aluminium crystals with a trace of holmium added. It is also used in high-strength magnets. In 2009, French scientists spectacularly claimed to have found holmium titanate crystals that behaved like monopoles (hypothetical particles that have only one magnetic pole – they are of interest to science geeks because the Nobel Prize-winning physicist Paul Dirac suggested that they must exist for the grand unified theory of physics to hold). The claim was heavily disputed, as the two poles of the crystals were really just extremely close together and not identical. In 2017, IBM made the more credible yet still astonishing claim that they had created a technique whereby a bit of data could be stored on a single holmium atom.

Erbium

In certain forms, erbium has very particular optical fluorescent properties, which are used in lasers. When it is added to the glass in fibre-optic cables, it amplifies the broadband signal being carried. It can also be alloyed to metals such as vanadium and used in infrared-absorbing glass, like several other lanthanides.

Thulium

The discovery of this lanthanide and that of holmium are both generally credited to Swedish scientist Per Teodor Cleve, although there were parallel investigations going on in several countries between 1878 and 1879. It is the second least common lanthanide (after promethium, which is self-destructing and only produced in nuclear reactions!). This means that, while it is not as rare as some other elements, it is expensive to produce, and as most of its properties are mirrored in cheaper lanthanides, it is not often used. However, one of its isotopes is employed in lightweight, portable X-ray machines, and it can also be used in surgical lasers.

Ytterbium

Sometimes described as the final element in the lanthanide sequence, ytterbium was identified in 1878 by Jean Charles Galissard de Marignac. It was discovered by heating erbium nitrate so that it decomposed into two oxides: erbium oxide and a white substance, mostly made up of a new element that he named ytterbium (although a pure sample was only created in 1953). It is mainly researched as a substitute or possible improvement on the uses of other lanthanides that it resembles. However, it might be used in the future to create even more accurate atomic clocks than those we already have – the isotope ytterbium-174 could in theory perform better than a caesium clock (which is already accurate to about a second every 100 million years!).

Lutetium

It eventually turned out that the sample of ytterbium produced by de Marignac was still not completely pure. The problem with the lanthanides is that their similarities make it extremely difficult to be sure you have isolated one. (The American chemist Theodore William Richards, in 1911, had to perform 15,000 successive recrystallizations of a sample of thulium bromate in order to isolate pure thulium with certainty.) In 1907, the French chemist Georges Urbain followed the same convoluted series of extractions that de Marignac had performed, and then showed that you could still extract a new element from the remaining sample of ytterbium: this was lutetium. Some chemists argue that the latter should be classed as a transition metal and included in the main body of the periodic table rather than as a lanthanide. The difficulty of extracting lutetium has meant that it is rarely used in isolation, although it has some commercial uses. For instance, it can be used in oil refineries as a catalyst for cracking hydrocarbons (in other words, breaking them down into simpler molecules).

Languid Centaurs

The lanthanides' chemical symbols are La, Ce, Pr, Nd, Pm, Sm, Eu, Gd, Tb, Dy, Ho, Er, Tm, Yb and Lu. Chemistry students struggling to remember this for an exam sometimes learn the following mnemonic: **La**nguid **C**entaurs **Pr**aise **N**e**d**'s **Pr**omise of **Sm**all **Eu**ropean **G**ar**d**en **T**u**b**s; **Dy**nosaurs **Ho**bble **Er**ratically **T**hru**m**ming **Y**e**b**low **Lu**tes.

Elements 72–94

Hafnium

72
Hf
178.49
Hafnium

Category: transition metal

Atomic number: 72

Colour: silvery grey

Melting point: 2,233°C (4,051°F)

Boiling point: 4,603°C (8,317°F)

First identified: 1923

To explain how hafnium was discovered, we need first to explain a significant breakthrough in our understanding of the periodic table. In 1911, the Dutch amateur physicist Antonius van den Broek suggested (without evidence) that the position of an element in the periodic table might better be defined by the amount of charge in the atom's nucleus. The young British physicist Henry Moseley had recently joined Ernest Rutherford's research group at the University of Manchester, where he created the world's first prototype atomic battery – he became fascinated by van den Broek's suggestion and set out to investigate it. He knew that when high-energy electrons collided with solids, they emitted X-rays. Returning to Oxford and funding his own research, he put together experimental apparatus that would allow him to fire electrons at various elements and measure the wavelengths and frequencies of the X-rays they emitted.

This led to the crucial discovery that each element emits X-rays at a unique frequency, and that this could be matched up perfectly to the element's atomic number (how many protons it has). This confirmed van den Broek's hypothesis and showed that the number of protons completely defined an element. Chemists quickly realized the significance of Moseley's work, as they could now rearrange the periodic table in a way that resolved any lingering doubts about its anomalies and flaws (and could also use X-rays as a much quicker way to identify elements).

Journey's End

Henry Moseley was strongly encouraged by his seniors to continue his scientific work after the Second World War broke out, but he insisted on joining the army. He was killed in 1915 at the Battle of Gallipoli. Robert Millikan, the physicist, wrote of him subsequently: 'a young man but twenty-six years old threw open the windows through which we can now glimpse the subatomic world with a definiteness and certainty never even dreamed of before. Had the European war had no other result than the snuffing out of this young life, that alone would make it one of the most irreparable crimes in history.'

This also meant that new gaps opened up in the table, prompting searches for the elements with sixty-one, seventy-two and seventy-five protons (and confirming that Mendeleev's final missing element 43 was still to be found).

Over the years, the elements 43, 61 and 75 would be shown to be technetium, promethium and rhenium.

Element 72 was discovered in 1923 by two young researchers, Dirk Coster and George de Hevesy, who worked at the institute of the great physicist Niels Bohr in Denmark. There had been ongoing debate about whether element 72 would turn out to be a lanthanide or a transition metal, but Bohr had argued it must be a metal. On this basis, Coster and de Hevesy investigated ores of zirconium, which was the transition metal above element 72 in the revised table. Within a few weeks, they had found traces of hafnium in the ores, using X-rays.

Hafnium has similar properties and uses to zirconium – both are used in nuclear reactors as they absorb neutrons well. It is also used in alloys that need to be strong and have a high melting point, and in plasma welding torches for the same reason.

Tantalum

73	
Ta	
180.94788	
Tantalum	

Category: transition metal

Atomic number: 73

Colour: blueish grey

Melting point: 3,017°C (5,463°F)

Boiling point: 5,458°C (9,856°F)

First identified: 1802

The Swedish chemist Anders Gustav Ekeberg identified tantalum in 1801, but there were many years of confusion over the similarity between tantalum and its upstairs neighbour niobium, before they were confirmed as separate elements. They almost always appear together in 'coltan', which is the name for any combination of columbite (a niobium-rich ore) and tantalite (which is rich in tantalum). Due to its stubborn resistance to reactions, it was named after the mythological Greek king Tantalus, who was punished for stealing from the gods by being forced to stand in a pool of water that would forever evade his attempts to drink it.

Tantalum is widely used in mobile phones and other handheld devices, such as games consoles and digital cameras. Capacitors, made of tantalum and its oxide, store up charge and electricity – the element is an excellent conductor of heat and electricity, so can be made into components that provide high capacitance at a very small size. It is unlikely that our current devices would be so miniaturized if it weren't for tantalum.

Among the metals, only tungsten and rhenium have higher melting points, so its alloys are used in hot places, such as aeroplane engines and nuclear reactors. And because it is so chemically inert, it has many medical applications – including in surgical instruments, implants such as pacemakers, and foil, gauze or wire for repairing nerves and muscles.

The high demand for tantalum in recent years has led to political controversy. Following the closure of a major mine in Australia during the economic downturn, the main source of the element now is the Democratic Republic of Congo, where profits from it helped to fund the dreadful civil war and continue to be tainted by corruption and political strife (leading some to denounce it as 'blood tantalum').

Tungsten

Category: transition metal

Atomic number: 74

Colour: silvery white

Melting point: 3,422°C (6,192°F)

Boiling point: 5,555°C (10,031°F)

First identified: 1783

I n the seventeenth century, Chinese porcelain makers used a tungsten pigment to create a lovely peach colour. Around the same time, in Europe, tin smelters would complain about how much their yield of tin fell when a certain ore was present – they called it 'wolf's foam', because it devoured the metal much as wolves devoured sheep – the name wolframite, for the mineral, probably derives from this usage.

After a few other scientists came close, the credit for tungsten's discovery is generally given to two Spanish brothers, the chemists Juan and Fausto Elhuyar. In 1783, they produced an acidic metal oxide and managed to reduce it to tungsten by heating it with carbon. They named it 'wolfram'.

It would come to be used for the filaments in incandescent lightbulbs, because it has the highest melting point of all metals. It is also used in quartz halogen lamps, where the addition of iodine allows it to be heated up even higher (creating a brighter light). Tungsten carbide is an extremely hard compound often used for drilling and cutting tools: it is used in mining and metal working as well as for high-performance dental drills. On a more prosaic level, biro (or ballpoint) pens have tungsten carbide 'balls' at the writing point.

You Say Tungsten, I Say Wolfram

The controversy over whether the metal should be called tungsten (the designation given to it by the Swedish chemist Carl Wilhelm Scheele, who studied the metal and named it after the Swedish words for 'heavy stone'), or wolfram, was theoretically settled in the early 1950s when the IUPAC ruled that 'columbium' should be called niobium and 'wolfram' tungsten. However, the name of wolfram is still reflected in the chemical symbol, and is occasionally still used, especially in Spain, where discontent lingers that the more poetic name chosen by the Elyuhar brothers has been abolished.

Rhenium

75
Re
186.207
Rhenium

Category: transition metal

Atomic number: 75

Colour: silvery

Melting point: 3,186°C (5,767°F)

Boiling point: 5,596°C (10,105°F)

First identified: 1925

There were a couple of false starts in the discovery of rhenium. In 1908, the Japanese chemist Masataka Ogawa isolated the element, which he called nipponium, but he wrongly announced that he had found element 43, so his findings were discredited. Rhenium was once again isolated in Germany in 1925 by Walter Noddack, Ida Tacke (who later married Noddack) and Otto Berg – they needed 660 kilograms of the ore molybdenite to produce a single gram of the metal (it is still produced as a by-product of the purification of copper and molybdenum) but also found it in gadolinite. They wrongly announced that they had found both elements 43 and 75, which harmed their reputation, but it was finally established that rhenium (named after the Rhine river) was indeed number 75. It was also the last naturally occurring metal to be discovered.

Rhenium is rare and usually found in nature with other metals, although some rhenium disulphide (a compound with sulphur) has been found at the mouth of a volcano in Eastern Russia. Rhenium dibromide is a super-hard material that, unlike diamonds, can be manufactured outside an extreme high-pressure environment.

Nuclear Error

The Noddacks' mistake in announcing they had found element 43 had one particularly unfortunate consequence. In 1934, Ida Noddack suggested that nuclear fission might be possible, but with her reputation having been damaged, the suggestion was ignored. The discovery of fission would be credited to Otto Hahn, Lise Meitner and Fritz Strassmann, who in 1938 became the first to recognize that the uranium atom, when bombarded by neutrons, actually split.

However, rhenium is mostly used, alloyed with nickel and iron, in the turbines of fighter aircraft. It is also a useful catalyst, which can be deployed in the manufacture of high-octane and lead-free petrol. And it can be used in alloys with tungsten and molybdenum, which are extremely hard and heat-resistant.

Osmium

Category: transition metal	**Melting point:** 3,033°C (5,491°F)
Atomic number: 76	**Boiling point:** 5,012°C (9,054°F)
Colour: blueish-silver	**First identified:** 1803

O smium was first identified in 1803 by the British chemist Smithson Tennant. Working with William Wollaston (see page 100), he melted crude platinum in *aqua regia* (a ferocious blend of acids used to melt metals) and it left a black residue. While Wollaston investigated the remaining platinum, Tennant experimented on this residue and discovered it could be separated out into two previously unknown metals: osmium and iridium. He was more impressed by iridium, naming osmium after the Greek word, *osme*, meaning smell, because it had a peculiar odour – several of its compounds also smell bad and the oxide is particularly unpleasant.

Depending on how you measure it, osmium is the densest element, about twice as dense as lead, but it has few contemporary commercial uses: only about 100 kilograms are currently produced each year. It is sometimes alloyed with iridium in expensive nibs for fountain pens, in surgical equipment, and in other tools that need be resistant to corrosion and wear. Osmium was one of the metals used in the filaments for lightbulbs, due to its high melting point, although tungsten came to be preferred for this purpose: the major German manufacturer of lighting equipment was named Osram in 1906 when both were in use – the name combines parts of the names of osmium and tungsten (bearing in mind that the German name for tungsten was still 'wolfram' at the time).

Iridium

77

Ir

192.217
Iridium

Category: transition metal

Atomic number: 77

Colour: silvery white

Melting point: 2,466°C (4,406°F)

Boiling point: 4,428°C (8,002°F)

First isolated: 1803

The metal iridium played an important role in confirming that the mass extinction of the dinosaurs (and other species) 65 million years ago was caused by a huge meteorite strike. It is rare on Earth but can often be found in meteorites. In 1980, the Nobel Prize-winning physicist Luis Alvarez and his colleagues from the University of California showed that there is an unusually high level of iridium in layers of the Earth's rock strata that were formed 65 million years ago – this has become known as the K–Pg boundary (because it is the meeting point of the Cretaceous and Paleogene eras, for which the abbreviations are 'K' and 'Pg').

This layer is visible at the surface in places, such as at Badlands in Alberta, Canada, or on the island of Zealand in

Denmark, but it is part of the fossil record around the world. Alvarez and his colleagues theorized that it was evidence of a massive meteorite impact at that time. The result was a long-lasting 'impact winter', which prevented plants from photosynthesizing and led to many species starving and disappearing. The theory was given greater weight in the 1990s with the discovery of the Chicxulub crater in the Gulf of Mexico – 180 kilometres wide; this impact crater confirmed that the K–Pg boundary was caused by debris being thrown up into the atmosphere and then settling after a huge meteorite strike.

The Goddess of the Rainbow

In Greek mythology, Iris, the goddess of the rainbow, was the daughter of the sea god Thaumas and his wife Electra. In between serving nectar to the gods and goddesses, and acting as their messenger, she is said to have gathered water from the ocean and used it to water the clouds with her pitcher. Because of the wide range of colourful salts it produces, Smithson Tennant named iridium after Iris, whose name is also the root of the word 'iridescent'.

In its pure metallic state, iridium is a brittle, shiny, silverish metal. As we've learned, it was first isolated by Smithson Tennant in 1803, along with osmium. For all the ingenuity of nineteenth-century scientists, it took decades for anyone to find a use for iridium, as the extremely high melting point made it difficult to work with. However, in 1834 the inventor John Isaac Hawkins, who wanted to create a thin, hard point for fountain-pen nibs, managed to create an iridium-pointed gold pen. Over time, methods were developed to work iridium and to create alloys of it with other metals – it is very hard and corrosion-resistant, and has been used to tip spark plugs, make aircraft parts and construct crucibles for use at very high temperatures. One of its isotopes, iridium-192, is also used in radiation therapy to treat patients with cancer.

Platinum

78
Pt
195.084
Platinum

Category: transition metal

Atomic number: 78

Colour: silvery white

Melting point: 1,768°C (3,215°F)

Boiling point: 3,825°C (6,917°F)

First identified: *circa* seventh century BC

An artefact made of worked platinum has been recovered from a casket dedicated to the Egyptian Queen Shapenapit – found at Thebes, it has been dated to the seventh century BC. South American civilizations were also working with the metal about 2,000 years ago. The Spanish conquistadors treated it with contempt, naming it 'platina' (little silver) and throwing it back into rivers on the assumption it was merely unripe gold.

However, a sample eventually made its way to Europe via a Spanish vessel captured by the British navy, and the metal started to gain a better reputation, although it would take a long time for anyone to develop cost-effective methods of producing it. It is a shiny metal, which is as resistant to corrosion as gold because it doesn't oxidize. One of the reasons platinum is so rare is that it is a heavy metal that can alloy with iron, so much of the platinum in the planet probably sank into the Earth's core in the past.

It now has a reputation as a prestige metal, used for wedding rings and giving its name to platinum records and weddings. It can be employed also in fuel cells, hard disks, thermocouples, optical fibres, spark plugs and pacemakers, among many other things. But its most significant use is in catalytic converters in cars, where it does a splendid job of transforming harmful hydrocarbons into carbon dioxide and water. This use is growing constantly and there is significant concern that global supply will be insufficient over coming decades.

Another important compound is cisplatin, which is the abbreviated name for cis-diamminedichloroplatinum(II). Barnett Rosenberg was experimenting with the effects of electrical currents on bacteria in the 1960s, and this compound was formed due to reactions with the electrodes; it turned out that it inhibited cell division among the bacteria. Cisplatin has become a valuable drug used to treat cancer of the testicles, ovaries and many other parts of the body.

Gold

79
Au
196.966569
Gold

Category: transition metal

Atomic number: 79

Colour: metallic yellow (or 'gold')

Melting point: 1,064°C (1,948°F)

Boiling point: 2,856°C (5,173°F)

First identified: ancient civilizations

Along with copper and silver, the elements above it in the periodic table, gold was known to the earliest civilizations. It was being used as jewellery and money at least 5,000 years ago. Gold can be found in nuggets (the largest ever found was in Australia in the 1860s, weighing in at over 70 kilograms) or in tinier pieces. You can, for instance, gather it from water by sieving rock – gold, being the heaviest metal, will always sink to the bottom. A large amount of gold was gathered using primitive methods. Tutankhamen's tomb contained over 100 kilograms of gold artefacts. The metal is chemically unreactive (although it will dissolve in *aqua regia*), soft enough to be cut with a knife and highly malleable, so it can be shaped with a hammer (24-carat gold is pure gold, while lower carats indicate alloys, which are slight harder).

Each year, we mine about 1,500 tons of gold (mostly from Russia and South Africa), and the global stock continues to be recycled and reused. Beaten into thin sheets, it can be used to electroplate other metals; for instance, in cheap gold jewellery or as a protection for electrical connectors. Computer chips often contain gold wires, which are used to make circuits. Gold is also used in alloys for dental fillings, and it is a catalyst in the production of PVA glue.

The Riches of the Ocean

After the First World War, Germany was required to make punitive reparations payments. The patriotic Nobel Prize-winning scientist, Fritz Haber, hatched an audacious plan to recoup the money by gathering particles of gold from seawater, using a combination of massive centrifuges and electrochemical methods. He estimated that a ton of seawater would yield 65 milligrams of metal particles, which would have made the plan economically viable. However, the true amount of gold in seawater is closer to 0.004 milligrams per ton, and when he recalculated this correctly, the scheme had to be abandoned, to his great sorrow.

The US Federal Reserve Bank in New York contains about 7,000 tons of gold bullion, belonging to various countries. It is worth something in the region of $500 billion and is the largest repository of gold in the world.

Mercury

80
Hg
200.592
Mercury

Category: transition metal

Atomic number: 80

Colour: greyish silver

Melting point: −39°C (−38°F)

Boiling point: 357°C (674°F)

First identified: ancient civilizations

The vivid red mineral cinnabar (also known as vermilion) was extensively traded millennia ago; in the Near East it was used as a rouge, and it had other colouring uses, such as in the amazing seventh-century Mayan Tomb of the Red Queen, which contains a sarcophagus and burial objects covered in bright red powder made from cinnabar. It was also well known that you could extract 'quicksilver' from cinnabar, a substance that 'dissolved' gold; theoretically, it could be used to extract the latter from other minerals – in particular, it would enable one to gather it more rapidly from river deposits in this way.

Except that the latter part wasn't quite right. Cinnabar is mercury sulphide, and the mercury can be extracted by heating it and collecting the evaporated metal. (The chemical symbol Hg comes from the Greek *hydrargyrum*, meaning liquid silver.) However, gold doesn't dissolve in liquid mercury – instead, the two metals amalgamate, forming a kind of compound, at an unusually low temperature. When you subsequently heat the mixture, the mercury evaporates leaving gold behind.

Mercury had a much better reputation in the past. The alchemists saw it as a primary form of matter upon which all other metals depended; the Romans and Greeks used it in medication and the Chinese believed that a mercury cocktail could guarantee long life.

Of course, we know now that mercury, the only metal which is liquid at room temperature, is toxic and that all these practices were sorely misguided. The Mad Hatter in *Alice's Adventures in Wonderland* is inspired by the mental derangement caused by the use of mercury nitrate in hat production. And one of mercury's most dangerous forms – methylmercury – can accumulate in fish, which will then make anyone who eats them very ill indeed.

Many former uses of mercury have been gradually phased out. In the past, you'd have found it in most thermometers, dental amalgams for fillings, fishing floats and paint pigments. It is still used in some chemical production methods today, but for the most part the allure that people felt towards this fascinating metal in the past has been replaced by great caution today.

Thallium

Category: post-transition metal	**Melting point:** 304°C (579°F)
Atomic number: 81	**Boiling point:** 1,473°C (2,683°F)
Colour: silvery white	**First identified:** 1861

Thallium is one of the most toxic elements and has been used in many murders over the years. It was the poison most often resorted to by Saddam Hussein to murder his opponents, and the 'thallium craze' in the early 1950s in Australia saw at least five separate crimes involving it. Until the 1970s, this deadly substance could easily be bought in the form of thallium sulphate, an insecticide and rat poison. It was discovered by William Crookes in 1861, who saw a thin green line in the spectrum of adulterated sulphuric acid, realized it must be a new element, and named it after the Greek word for a green shoot or twig: *thallos.*

In 1862, the French scientist Claude-Auguste Lamy did some more detailed research and purified a small amount of the soft, silver metal (which tarnishes rapidly in the air); there followed a furious controversy between Crookes and Lamy over who should get the credit for this, only settled when each was awarded their own medal.

Thallium is mainly found in ores such as potassium minerals and pollucite (with caesium). It is the similarity to potassium that makes it so dangerous. It can hijack the parts of cells that require potassium, and this interferes with the important roles played by the latter.

In the short term, thallium poisoning causes nausea and diarrhoea; over longer periods, it causes extensive nerve damage, hair loss, mental disturbances and heart failure. Strangely, the most effective antidote is potassium 'ferrihexacyanoferrate' (known as Prussian blue or Berlin blue), a substance that contains cyanide, but in a non-toxic form. It works by surrounding the thallium molecules and preventing them from being absorbed in the place of potassium.

Some thallium is produced as a by-product of copper and lead refining – it has few uses, other than in the electronics industry, as part of photoelectric cells, and in the form of thallium oxide to create glass with a low melting point.

Lead

Category: post-transition metal

Atomic number: 82

Colour: dull grey

Melting point: 327°C (621°F)

Boiling point: 1,749°C (3,180°F)

First identified: ancient civilizations

The alchemists regarded heavy, malleable lead as a lowly metal, but knew that it could be turned from its natural grey to a variety of other colours. Soaked in vinegar and left in a shed of animal dung, it would turn white. Heated, it would form a surface layer of yellow lead monoxide, which they called 'litharge', and then a bright red (which was used as a red paint in the Middle Ages – although this fades over time to a dull brown). Some wrongly believed that if they kept working on the metal, it might eventually turn to gold.

It was obtained from the mineral galena at least as long ago as the Ancient Greeks. The Romans used it for piping, pewter, paint, pottery glazes and even cosmetics (in the form of lead carbonate or 'lead white', which was also used as a paint pigment), although the doctor Cornelius Celsus warned against the bad effects of this.

Trust Me, I'm a Scientist

In the US, in 1924, a press conference was held after an outbreak of lead poisoning at the Standard Oil plant in New Jersey – one worker had become psychotic and died, and thirty-five more were hospitalized. Thomas Midgley, the inventor of lead petrol, had himself only just recovered from lead poisoning in Florida, though he tried to convince sceptical reporters of the fuel's safety by washing his hands in a container of the additive tetraethyl lead, as well as claiming that the petrol would be safe because 'the average street will probably be so free from lead that it will be impossible to detect it or its absorption'. He did, however, concede that 'no actual experimental data has been taken'.

Today, in spite of its toxicity, it is still used in car batteries, some pigments, weights and in solders. As the heaviest stable element that is not radioactive, it can be used for radiation protection; for instance, in containers that contain mildly radioactive material. Lead is not especially reactive, so can be used to hold corrosive acids. Until recently it was used to prevent knocking (ignition problems) in car engines, but this has been banned due to the pollution it caused. It is no longer used in water pipes and containers, but still causes some nasty cases of lead poisoning in antiquated buildings that contain lead piping.

Incidentally, the alchemists weren't completely wrong. Many radioactive elements with a higher atomic number than 82 turn into lead at the end of their decay chains, so, in theory, it is easier to turn gold into lead than vice versa. Nuclear experiments have shown that it is feasible to make the opposite transmutation, but the expense would far outweigh the possible gain.

Bismuth

83
Bi
208.98040
Bismuth

Category: post-transition metal

Atomic number: 83

Colour: pinkish silver

Melting point: 272°C (521°F)

Boiling point: 1,564°C (2,847°F)

First identified: fifteenth century

B ismuth was known to the Incas in the fifteenth century; a knife found at Machu Picchu is made of an alloy containing the metal. Western alchemists had also recognized bismuth, and it was being mined by 1460, although it was often mistaken for a type of lead. In the nineteenth century, it was used in cosmetics: when dissolved with nitric acid and then poured into water, it produces a white, flaky material known as 'pearl white', which could be made into a face powder. This is much less toxic than lead white, but tended to turn rather brown in cities, due to sulphur pollution from coal burning.

Bismuth is a heavy but brittle metal, often used in alloys such as pewter. With cadmium or tin, it forms alloys with a low melting point, which can be used in fuses or solders. It is still used (as bismuth chloride oxide) to create a pearly effect in cosmetics and (as bismuth oxide) for yellow pigments. Bismuth carbonate is sometimes used as an indigestion cure (known as 'bismuth mixture').

A Brief Guide to Radioactivity

In the nuclei of stable atoms, there is enough force to bind the protons and neutrons together. In unstable atoms, though, especially heavy ones such as uranium, the force is not strong enough, so the nucleus emits energy and particles, which we call 'radioactive decay'. (Note that some usually stable elements do have radioactive isotopes.) The term radioactivity refers to the particles emitted: the atom will gradually decay until it becomes stable. For instance, uranium-238 decays gradually through eighteen stages, forming atoms of thorium, radium, radon and polonium, before becoming a stable atom of lead-206. It's impossible to estimate how long a single atom will take to decay, so instead we use the concept of 'half-life', which is the average amount of time taken for half of the nuclei of a given isotope to decay.

It used to be thought that bismuth wasn't radioactive. In fact, it is, but only very, very slightly. In 2003, a group of researchers in France detected alpha particles resulting from the decay of bismuth-209 (the only naturally occurring isotope of bismuth). However, it has a half-life of 2×10^{19}: only a handful of substances have longer half-lives, so its radioactivity is not at all dangerous, unlike most of the substances beyond it in the periodic table.

Polonium

Category: metalloid
Atomic number: 84
Colour: silvery grey
Melting point: 254°C
(489°F)

Boiling point: 962°C
(1,764°F)
First identified: 1898

M arie Curie didn't discover radioactivity – X-rays had been identified in 1895 by Wilhelm Röntgen, and uranium radiation in 1896 by Henri Becquerel, who she would go on to work with. However, she did coin the term and did a great deal, together with her husband Pierre, to expand our understanding of this phenomenon.

The extraction of polonium by Marie and Pierre was a tricky undertaking. They were exploring the radioactive ore pitchblende (now known as uranite), which contains uranium but seemed to be more radioactive than it should be on account of that element. They managed to remove the uranium and sieved through tons of the remaining rubble to find a few crumbs of polonium, named after Marie's native Poland.

Polonium is extremely rare, and it is uneconomical to use Curie's method to extract it. Instead, bismuth-209 is bombarded with neutrons, which creates bismuth-210, which in turn decays to form polonium. It can be classed as a metal rather than metalloid, as its electrical conductivity falls at high temperatures – a quality that means it can eliminate static electricity in some industrial processes. It has a short half-life, which means it generates a lot of heat. This has been exploited as a way to generate thermoelectric power in satellites and lunar vehicles, such as the Russian 'Lunokhod' rovers that explored the moon's surface.

Murder by Polonium

The murder of the Russian ex-agent Alexander Litvinenko in 2006 was carried out using a small amount of polonium – it emits alpha particles (containing protons and neutrons) that are weakly penetrating, so it is a relatively safe substance to (for instance) carry in a small container. However, if polonium is ingested, the same radiation becomes extremely dangerous as it attacks cells within the body by being absorbed into them, and this was what killed the unfortunate Litvinenko.

Astatine

Category: halogen
Atomic number: 85
Colour: n/a
Melting point: 302°C (576°F)

Boiling point: 337°C (639°F)
First identified: 1940

milio Segrè, the co-discoverer of the first 'artificial' element technetium in 1937, spent the following summer at Berkeley; and when Italy passed anti-Semitic laws that would have barred him from being a professor, he chose to stay there, and ended up using the particle accelerator at Berkeley to discover another new element, 'astatine', whose name is derived from *astatos*, the Greek word for 'unstable'.

It only occurs naturally as part of a complex radioactive pathway. It has ten highly radioactive isotopes, none of which has a half-life of more than eight hours. To create a tiny quantity, Segrè and Dale Corson bombarded bismuth-209 with particles to form astatine-211, albeit not in a quantity large enough to be seen. It is one of the halogens, and probably has similar properties to them.

Segrè went on to work at the Manhattan Project, and there has not been much work done on astatine since. It is thought, however, that it could be used to treat some kinds of cancer. The radioactive isotope iodine-131 has been used for this, but has the disadvantage of emitting beta particles (high-energy electrons), which can damage other tissue outside of the tumour. Astatine-211 is an alpha emitter with a very short half-life, which might make it a better option in future.

Radon

Category: noble gas
Atomic number: 86
Colour: glowing orangey red in solid form, otherwise colourless

Melting point: −71°C (−96°F)
Boiling point: −62°C (−79°F)
First identified: 1900

You may be alarmed to hear that there is a colourless, odourless, radioactive gas that is constantly seeping from the ground, and that can build up in dangerous quantities in poorly ventilated basements (especially in granite buildings). Radon is the first of two radioactive noble gases that complete the bottom right-hand corner of the periodic table. It is formed through the decay of small quantities of uranium in the soil, in a pathway also including radium, thorium and actinium. In turn, it has a short half-life and rapidly decays into polonium, then bismuth, then lead.

Radon can be collected by placing a glass jar over a piece of radium. It was first described by the German chemist Friedrich Ernst Dorn as a 'radium emanation' – a gas that appeared to make the air around a piece of radium radioactive. Ernest Rutherford and William Ramsay subsequently had a slightly bitchy dispute over who was the element's true discoverer.

Its short half-life means that when radon does gather in a building, it will gradually disappear, but there have been cases in which it can be quite damaging – it can, for instance, contribute to causing lung cancers. There are home testing kits available if you are worried about your own basement. However, most radon passes fairly harmlessly into the atmosphere, of which it forms a minute proportion while it decays.

Francium

Category: alkali metal
Atomic number: 87
Colour: unknown
Melting point: 21°C (70°F)

Boiling point: 650°C (1,202°F)
First identified: 1939

In 1929, five years before her death, caused by radiation exposure, Marie Curie hired a new lab assistant, Marguerite Catherine Perey. The brilliant Perey discovered the elusive element francium (named after France) in 1939, and would go on to be the first woman to be elected a member of the French Academy of Sciences.

When a particle loses an alpha particle, its atomic number falls by two; when it emits a beta particle, its atomic number rises by one. When it comes to artificially creating radioactive elements, the trick is to consider the possible decay pathways from other elements. In the case of francium,

Perey purified a sample of actinium (atomic number 89) of its known radioactive impurities, but found a remaining trace of radioactivity, which turned out to be the new element.

Speedy Electrons

We have seen that the other alkali metals (lithium, sodium, potassium, rubidium and caesium) become more reactive as you go down the periodic table. Francium is an exception for an interesting reason – as the atoms of the elements have more and more protons, the electrons move around at an incredible pace, approaching the speed of light. The laws of relativity mean that they are thus slightly smaller than they would be at lower speeds, and the electrons are packed closer to the nucleus and fractionally harder to remove. As a result, caesium has been shown to be more reactive than francium (although you wouldn't want to drop a lump of either of them into your bath).

Most actinium decays by emitting a beta particle, to form thorium (with an atomic number of 90, which in turn loses an alpha particle to become radium with an atomic number of 88). However, a tiny proportion of actinium atoms lose an alpha particle instead and become element 87, francium, which can thus only ever be found naturally in tiny quantities and has a very short half-life.

Radium

Category: alkaline earth metal	**Melting point:** 700°C (1,292°F)
Atomic number: 88	**Boiling point:** 1,737°C (3,159°F)
Colour: whiteish	**First identified:** 1898

When the Curies found polonium in pitchblende, they also found radium (which was named for its property of glowing in the dark). In 1911, Marie Curie went on to isolate the metal (with her colleague André Debierne) by electrolysing radium chloride with a mercury cathode.

The Radioactive Cookbook

It was very possibly radium that killed Marie Curie (she died of aplastic anaemia) – when she discovered radium, she used to enjoy going to the lab in the dark and watching the test-tubes glowing like fairy lights. The notebooks and papers she left behind still have to be kept in lead boxes and can only be viewed using radiation protection. Even the cookbook in her kitchen turned out to be extremely radioactive, presumably because she had handled it.

It occurs naturally in small quantities within uranium ores. A highly radioactive element, it has had some medical applications, notably in early cancer treatments, but these are mostly now redundant, although radium-223 is sometimes used to treat prostate cancer when it has spread to nearby bones.

Early in the twentieth century, radium was used in small quantities in luminous paints such as those used on clock dials. A famous court case of the 1920s involved the 'radium girls', five young women who had developed tumours after working in the US radium factory – they were using paint containing radium and were issued with no safety guidance. As well as handling the paint, some used to lick the tip of their paintbrushes into a point, thus ingesting small quantities of it. They won the case, but all died within a few years. Radium is no longer used in luminous paint, partly as a result of their courage in pursuing and winning that court case.

Actinium

Category: actinide
Atomic number: 89
Colour: silver
Melting point:
1,050°C (1,922°F)

Boiling point: 3,200°C
(5,792°F)
First identified: 1899

L ike the lanthanides, the actinides are a series of elements (from actinium, element 89, to lawrencium, element 103) that are displayed in a strip outside of the periodic table. However, the actinides show greater diversity in their properties and some are highly significant, especially uranium, so let's go into a bit more detail about the first few actinides, at least up until plutonium (element 94), after which we find ourselves ever deeper into the realms of artificially created elements.

The actinides do share some common properties: all are radioactive in each of their isotopes; they all tarnish in air and spontaneously ignite in air (especially as a powder); they all react with hot water and release hydrogen; and they are all soft, dense, silver metals.

Actinium was discovered by Marie Curie's friend André Debierne, using the same method they used to find radium. Pitchblende (or uranite), the uranium ore from which it is gathered, tends to glow, giving off an eerie, blue light. This is largely down to the actinium content. The amount of actinium in pitchblende is tiny, so when it is needed for research purposes (it has few practical uses other than in some smoke detectors and experimental radiotherapy), it is manufactured by bombarding radium-226 with neutrons.

Thorium

Category: actinide
Atomic number: 90
Colour: silver
Melting point: 1,750°C (3,182°F)

Boiling point: 4,788°C (8,650°F)
First identified: 1829

The streets of many cities used to be lit up with a radioactive element: thorium oxide has the highest melting point of all oxides, and this led to it being used in the gaslights of the late nineteenth and early twentieth century. In the heat of the burning gas, it didn't melt but instead gave off a bright white light. Fortunately, thorium is not as radioactive as some of its fellow actinides, and emits alpha particles, which would not penetrate through glass or human skin, so this was a safer method of lighting than it sounds. Indeed, it is still used in some camping equipment, although you will generally find versions that are specifically labelled as 'thorium-free'!

Thorium is relatively abundant – there is three times as much of it in the Earth's crust as uranium. This is because, while it is part of various radioactive decay chains, its half-life in the naturally occurring isotope thorium-232 is longer than the age of our planet.

Discovered by Jöns Jacob Berzelius in 1828, thorium was named after the Viking thunder god, Thor. Of course, he didn't realize it was radioactive – that concept was unknown at the time. It is sometimes used in nuclear reactors instead of uranium. As thorium and uranium are not always found in the same places, some countries are working to build up their thorium reactors. For instance, India, whose east coast is rich in monazite (a thorium source), has been developing new technology that will enable it to use thorium more efficiently in the future.

Protactinium

91	
Pa	
231.03588	
Protactinium	

Category: actinide
Atomic number: 91
Colour: silver
Melting point: 1,568°C (2,854°F)

Boiling point: 4,027°C (7,280°F)
First identified: 1913

Protactinium has had a few different names over the years. In 1900, the English scientist William Crookes noted that there was an unknown radioactive substance in some uranium ores: he called it uranium-X. In 1913, the Polish–American chemist Kasimir Fajans isolated the isotope protactinium-234, which he named 'brevium' due to its half-life of about a minute. However, when the German physicist Lise Meitner isolated a different isotope, protactinium-231, with a half-life of 33,000 years, Fajans suggested renaming the element. Meitner called it 'protoactinium', because the element forms actinium when it decays, losing an alpha particle in the process. This was a bit awkward to pronounce, so eventually it was shortened to protactinium.

There are few practical applications for this scarce element (which is hard to refine). However, it can be used to reconstruct the movements of bodies of water in the ocean by comparing the proportions of protactinium-231 and thorium-230. Both are present in small amounts due to the decay of uranium particles in the sea. But the thorium decays faster than the protactinium, so researchers can use the ratio between the two to model the circulation of water.

Uranium

Category: actinide
Atomic number: 92
Colour: silvery grey
Melting point: 1,132°C (2,070°F)

Boiling point: 4,131°C (7,468°F)
First identified: 1789

Pitchblende (or uranite), the ore from which several other radioactive elements would be extracted, was known to medieval silver miners, who occasionally unearthed the black or brown mineral. Martin Klaproth studied it in

1789 and managed to produce a yellow compound that he correctly believed to contain a new element (which he named after the planet Uranus). Uranium was later isolated by the French chemist Eugène Péligot. It was not known that it was radioactive until 1896, when Henri Becquerel left a sample on an unexposed photographic plate, which became cloudy, suggesting that it must be giving off some kind of rays.

Most of the uranium in the planet is uranium-238 (about 99 per cent) while a small proportion is uranium-235, and there are tiny quantities of a few other isotopes. Uranium has a long half-life, which is why there is quite a large amount of it still available – the radioactive decay of the mineral is a significant source of heat inside the planet (and drives phenomena such as volcanoes). It has also played a significant role in allowing us to estimate the planet's age – it is formed in supernovae in a ratio of uranium-235 to uranium-238 of about 8:5. Comparing the original amounts of each to their current ratio (and taking their half-lives into account) gives us an estimate of the Earth's age.

Uranium is the only naturally occurring element that can be used as fuel in nuclear reactors. It is also used to power nuclear submarines and (along with plutonium) to build nuclear weapons, which work by releasing huge amounts of energy and radiation as atoms are either torn apart (nuclear fission) or fused together (nuclear fusion) in a detonation.

Neptunium

Category: actinide
Atomic number: 93
Colour: silver
Melting point: 637°C
(1,179°F)

Boiling point: 4,000°C
(7,232°F)
First identified: 1940

The Italian-American physicist Enrico Fermi tried to create elements 93 and 94 by bombarding thorium and uranium with neutrons. He believed he had succeeded, but it was later shown that he had accidentally discovered nuclear fission and was finding fission products of the original elements. Fermi would go on to work on the Manhattan Project and to build the world's first nuclear reactor, the Chicago Pile-1.

In 1940, Edwin McMillan and Philip Abelson at Berkeley used Fermi's method and succeeded in creating element 93, which they named after the planet Neptune, as it is next to Uranus in the solar system. Neptunium is the last naturally occurring element, found in uranium ores in trace amounts. It is also present in vanishingly small quantities in many houses, because the radioactive element americium, which is used in tiny quantities in some smoke detectors, decays to form it.

Plutonium

| 94 |
| Pu |
| 244 |
| Plutonium |

Category: actinide

Atomic number: 94

Colour: silvery white

Melting point: 639°C (1,183°F)

Boiling point: 3,228°C (5,842°F)

First identified: 1940

Plutonium was discovered in the same year as neptunium, also at Berkeley: neptunium was synthesized and this decayed, losing a beta particle to form plutonium-239 (which in turn decays to form uranium-235). It was named after Pluto, since the previous two elements had also been named after planets.

It is quite an interesting metal – at room temperature it is brittle, but if you heat it up or alloy it with gallium, it becomes more malleable and workable. It can also be alloyed with cobalt and gallium to produce a material that is a superconductor at low temperatures (although this doesn't last long, as the plutonium rapidly decays, damaging the material in the process). Plutonium-238 also used to be employed as a thermoelectric generator in old pacemakers. And its ability to generate heat as it decays has allowed scientists to create power sources in probes such as *Cassini*, which explored Saturn.

The Plutonium Prank

The team of scientists that synthesized what came to be known as plutonium during the A-bomb project in the Second World War was led in the US by Glenn Seaborg – apparently a guy with a quirky sense of humour. In the short term, the A-bomb project was top secret and so the codename 'copper' (while referring to real copper as 'honest-to-God copper') was used for the element. When the war ended, and Seaborg was finally allowed to rename it, he spurned the obvious choice of 'Pl' for the abbreviation, instead choosing 'Pu' – 'he just thought it would be fun' to name it 'Pee-Yu', because that is what a child might say when they encounter something really stinky. Seaborg's joke was approved by the naming committee and is now immortalized in the periodic table.

Of course, plutonium is best known for being one of the elements used to power nuclear weapons – 'Little Boy' (the bomb that was used on Hiroshima) was a uranium weapon, but 'Fat Man' (which subsequently devastated Nagasaki) used plutonium to horrible effect.

Elements 95–118

N eptunium is the last naturally occurring element, while plutonium, which is created in supernovae (and by irradiating uranium), has played a highly significant role in our history. From this point onwards, the elements become increasingly arcane – they can only be created on Earth by bombarding other elements with particles in a handful of high-tech laboratories around the world, and they are all highly unstable and rapidly decay back into uranium and other elements. So, rather than devote a whole section to each of them, here are a few key facts, with the chemical symbols and atomic numbers in brackets.

Americium

Americium (Am 95) was once present on Earth, having been formed in the natural nuclear reactions beneath Gabon, but the longest-lived isotope (americium-247) has a half-life of 7,370 years, meaning that supply has all decayed. It was first manufactured in 1944 at the University of Chicago by a team led by Glenn Seaborg.

Curium

Curium (Cm 96), which was named after the Curies, had also been discovered by a group led by Glenn Seaborg, this time at Berkeley earlier in 1944. Seaborg actually made the announcement when he appeared on a children's radio show in November 1945. It has been used as a power source on missions to space.

Berkelium

Berkelium (Bk 97) was made by bombarding americium-241 with helium particles in 1949. It took nine years to make enough of the element to be seen with the naked eye. It was named after Berkeley, where it was created.

Californium

Californium (Cf 98) continued the theme, being named after the state where it was created by bombarding curium atoms with helium. It is used in detectors that identify gold and silver ores, and for detecting metal fatigue in aeroplanes.

Einsteinium and Fermium

Einsteinium (Es 99) and Fermium (Fm 100) were discovered in fall-out material from the nuclear test in November 1952 at Bikini Atoll. Both were initially kept secret and only announced as new elements in 1955. The elements with a higher atomic number than 100 are called the 'transfermium elements'.

Mendelevium

Mendelevium (Md 101) is rather splendidly named after the creator of the periodic table. When it was first created in the Berkeley cyclotron, just seventeen atoms were produced. Like most of the heavier elements, it is only used for research purposes.

Nobelium

Nobelium (No 102) was the cause of a scientific spat. It was discovered, but not announced, in 1956 by scientists at the Institute of Atomic Energy, Moscow. After it was subsequently made at the Nobel Institute in Stockholm (thus the name) and also in Berkeley, years of debate followed as to who the real discoverers were.

Lawrencium, Rutherfordium and Dubnium

Lawrencium (Lr 103), Rutherfordium (Rf 104), Dubnium (Db 105) – Russian and American teams also squabbled over who discovered elements 103, 104 and 105. Element 103 is named after Ernest Lawrence, who invented the cyclotron particle accelerator. The Russians first made the element now known as Rutherfordium (named after the physicist Ernest Rutherford) in 1964 by bombarding plutonium with neon. A similar technique led to the discovery of element 105, which the Russians called neilsbohrium, and the Americans hahnium. The IUPAC eventually ruled it should be called dubnium after Dubna, home of the Russian Joint

Institute for Nuclear Research (JINR). Lawrencium is the last actinide: elements from 104 upwards can be called transactinides or super-heavy elements.

Seaborgium

Seaborgium (Sg 106), named after Glenn Seaborg, was first made in 1970 by bombarding californium with oxygen, and again in 1974 by bombarding lead with chromium. Only a small number of atoms have ever been produced.

Bohrium, Hassium, Meitnerium, Darmstadtium, Roentgenium and Copernicum

Bohrium (Bh 107) was probably first made at the JINR in 1975, but the first sighting production was at the German nuclear research institute (the Gesellschaft für Schwerionenforschung or GSI) after bismuth was bombarded with chromium in a cold fusion process (meaning it happened at or near to room temperature). The same team also produced the first tiny amounts of Hassium (Hs 108), Meitnerium (Mt 109), Darmstadtium (Ds 110), Roentgenium (Rg 111) and Copernicum (Cn 112).

Nihonium

Nihonium (Nh 113) was produced by scientists from RIKEN (the Institute of Physical and Chemical Research) in Japan in 2004, and very little is known about it to this day.

Flerovium, Moscovium, Livermorium, Tennessine and Oganesson

The last five elements were all created at the JINR by Yuri Oganessian and his team of scientists. They are Flerovium (Fl 114), Moscovium (Mc 115), Livermorium (Lv 116), Tennessine (Ts 117) and Oganesson (Og 118, named after the man himself). Tennessine was the last of the 118 elements to be synthesized, in 2010. All of these elements are highly unstable and have only been synthesized in tiny quantities, so we know relatively little about them.

Element 119 and Beyond

I f you are a *Star Trek* fan, you may know of dilithium, the crystalline element (with an atomic number of 119) that powers future space vessels, having been discovered on a moon of Jupiter or in a meteorite at the South Pole (depending on which episode you watch). Of course, the writers made that up (as did the writers of the *Batman* comic, whose plot relied on the discovery of element 206, batmanium) – but the search to find the real element 119 and other super-super-heavy elements really is underway. These will be extraordinarily difficult to create, as they will involve bombarding atoms with particles, possibly for years, to find just a few atoms of the new element (which will be tremendously unstable and will rapidly decay). The Japanese RIKEN team, in collaboration with the Oak Ridge National Laboratory (ORNL) in Tennessee, believe they can possibly do it by bombarding curium with vanadium ions. Yuri Oganessian's team in Russia are planning an attempt using berkelium, which they will bombard with titanium ions.

Carl Sagan once said that 'we are made of star stuff', expressing the sheer wonder of how all of the elements were created in the Big Bang, or in nuclear reactions in stars and supernovae, then swirled around in space and came together to form everything in our world, both living and inorganic, including every molecule in our bodies.

Before 1669, we only knew of twelve elements. By the end of the eighteenth century we knew of thirty-four. Mendeleev's periodic table included the sixty-two elements that were known at the time. And now we know about each of the first 118, including all naturally occurring elements on our planet, and we are still not satisfied, because of the natural human inclination to keep on trying to do more, go further, and reach out for the stars.

Index

(*n* refers to footnotes)

orpiment 79
Ørsted, Hans Christian 40
Osann, Gottfried 99
osmium (Os) 141–2
Osram 142
oxy-gène 28
oxygen (O) 5, 27–9, 54, 118, 124
ozone 29, 50

P

Pacific Coast Borax Company 18
palladium (Pd) 100, 102–3
Paneth, F.A. 97
Paracelsus 8, 114
Paris Exhibition (1910) 33
Péligot, Eugène 168
Perey, Marguerite Catherine 160, 161
periodic law, creation of 2
periodic table, discovery of 1
Perrier, Carlo 97
philosopher's stone 44
phlogisticated air 51
phlogiston 8, 9, 28
phosphorous (P) 5, 44–6, 127–8
Pilkington's ActivT 59
pitchblende 156, 164, 167
platinum (Pt) 99, 100, 117, 141, 144–5
Plunkett, Roy 31
Plutonium Prank 171
plutonium (Pu) 97, 164, 170–71, 172
polonium (Po) 77n, 156–7, 159
Pons, Stanley 103

potassium 'ferrihexacyanoferrate' 151
potassium (K) 19, 34, 53–5, 118, 161
praseodymium (Pr) 123, 124, 125–6, 132
Priestley, Joseph 28
promethium (Pm) 123, 126–7, 132, 135
protactinium (Pa) 166–7
protium 3
Proust's law 93

Q

quartz 41, 42
quicksilver 148

R

radioactivity, brief guide to 155
radium (Ra) 162–3
radon (Rn) 159–60
Ramsay, Sir William 12, 32, 33, 51, 52, 84, 116, 159
rare earths 122–32
Reich, Ferdinand 108
Reichenstein, Franz Joseph Müller von 113
Respour, P. Moras de 72
rhenium (Re) 135, 139–40
rhodium (Rh) 100–101, 102, 141
Richards, Theodore William 132
Richter, Hieronymus 109
Robards, Marie 120
roentgenium (Rg) 175

tennessine (Ts) 176
terbium (Tb) 123, 128–9, 132
thallium (Tl) 150–51
Thénard, Louis-Jacques 19
thorium (Th) 159, 165–6
thulium (Tm) 123, 128, 131, 132
tin (Sn) 109–10
titanium (Ti) 58–9, 177
Tokyo subway (1995) 45
transfermium elements 173
Travers, Morris 32, 84, 116
tritium 3
The Triumphal Chariot of Antimony (Valentine) 112
tungsten (W) 137–8, 142
20 Mule Team Borax 18–19

U

uranium (U) 16, 126, 156, 159, 163, 164, 166, 167–8, 169, 172
Uranus 113, 114, 168
Urbain, Georges 132

V

Valentine, Basil 111–12
vanadium (V) 15, 60–61
Vauquelin, Louis Nicolas 16, 61

Vinci, Leonardo da 28

W

Weintraub, Ezekiel 19
Welsbach, Carl Auer von 125
witherite 121
Wöhler, Friedrich 39, 90
wolfram 137, 138, 142
Wollaston, William 100, 141

X

X-rays 16, 120, 129, 131, 133–4, 135, 156, 166
xenon (Xe) 33, 116–17
XIPS 117

Y

YBCO 90–91, 121
ytterbium (Yb) 123, 128, 131, 132
yttrium (Y) 89–91, 119, 122, 129, 130

Z

zeppelins 9
zinc (Zn) 72–3, 104, 106, 108
zincblende 74
zirconium (Zr) 91–2, 135

THE PERIODIC TABLE OF THE ELEMENTS

Group	Element	Atomic No.	Name	Atomic Weight
1 IA	H	1	Hydrogen	1.008
1 IA	Li	3	Lithium	6.94
2 IIA	Be	4	Beryllium	9.0122
1 IA	Na	11	Sodium	22.990
2 IIA	Mg	12	Magnesium	24.305
1 IA	K	19	Potassium	39.098
2 IIA	Ca	20	Calcium	40.078
3 IIIB	Sc	21	Scandium	44.956
4 IVB	Ti	22	Titanium	47.867
5 VB	V	23	Vanadium	50.942
6 VIB	Cr	24	Chromium	51.996
7 VIIB	Mn	25	Manganese	54.938
8 VIIIB	Fe	26	Iron	55.845
VIII	Co	27	Cobalt	58.933
1 IA	Rb	37	Rubidium	85.468
2 IIA	Sr	38	Strontium	87.62
3 IIIB	Y	39	Yttrium	88.906
4 IVB	Zr	40	Zirconium	91.224
5 VB	Nb	41	Niobium	92.906
6 VIB	Mo	42	Molybdenum	95.95
7 VIIB	Tc	43	Technetium	(98)
8 VIIIB	Ru	44	Ruthenium	101.07
VIII	Rh	45	Rhodium	102.91
1 IA	Cs	55	Caesium	132.91
2 IIA	Ba	56	Barium	137.33
3 IIIB	57–71		Lanthanides	
4 IVB	Hf	72	Hafnium	178.49
5 VB	Ta	73	Tantalum	180.95
6 VIB	W	74	Tungsten	183.84
7 VIIB	Re	75	Rhenium	186.21
8 VIIIB	Os	76	Osmium	190.23
VIII	Ir	77	Iridium	192.22
1 IA	Fr	87	Francium	(223)
2 IIA	Ra	88	Radium	(226)
3 IIIB	89–103		Actinides	
4 IVB	Rf	104	Rutherfordium	(267)
5 VB	Db	105	Dubnium	(268)
6 VIB	Sg	106	Seaborgium	(269)
7 VIIB	Bh	107	Bohrium	(270)
8 VIIIB	Hs	108	Hassium	(270)
VIII	Mt	109	Meitnerium	(278)

Lanthanides

Element	Atomic No.	Name	Atomic Weight
La	57	Lanthanum	138.91
Ce	58	Cerium	140.12
Pr	59	Praseodymium	140.91
Nd	60	Neodymium	144.24
Pm	61	Promethium	(145)
Sm	62	Samarium	150.36
Eu	63	Europium	151.96

Actinides

Element	Atomic No.	Name	Atomic Weight
Ac	89	Actinium	(227)
Th	90	Thorium	232.04
Pa	91	Protactinium	231.04
U	92	Uranium	238.03
Np	93	Neptunium	(237)
Pu	94	Plutonium	(244)
Am	95	Americium	(243)